图书在版编目（CIP）数据

疗养院与康复中心设计 /（美）卡尔编；常文心，张晨译. — 沈阳：辽宁科学技术出版社，2014.7
ISBN 978-7-5381-8638-3

Ⅰ. ①疗… Ⅱ. ①卡… ②常… ③张… Ⅲ. ①疗养院－建筑设计 ②康复医院－建筑设计 Ⅳ. ①TU246.2

中国版本图书馆CIP数据核字（2014）第103080号

出版发行：辽宁科学技术出版社
（地址：沈阳市和平区十一纬路29号 邮编：110003）
印　刷　者：利丰雅高印刷（深圳）有限公司
经　销　者：各地新华书店
幅面尺寸：215mm×285mm
印　　张：13
插　　页：4
字　　数：50千字
印　　数：1～1500
出版时间：2014年 8 月第 1 版
印刷时间：2014年 8 月第 1 次印刷
责任编辑：陈慈良　殷　倩
封面设计：杨春玲
版式设计：杨春玲
责任校对：周　文
书　　号：ISBN 978-7-5381-8638-3
定　　价：218.00元

联系电话：024-23284360
邮购热线：024-23284502
E-mail: lnkjc@126.com
http://www.lnkj.com.cn
本书网址：www.lnkj.cn/uri.sh/8638

DESIGN OF
NURSING HOMES AND
REHABILITATION FACILITIES

疗养院与康复中心设计

（美）罗伯特·F·卡尔／编　　常文心　张晨／译
Edited by Robert F. Carr　Translated by Catherine Chang, Zhang Chen

辽宁科学技术出版社

CONTENTS
目录

Nursing Home — 006 疗养院
- Overview — 006 概述
- Building Attributes — 006 建筑属性

TO BUILD UP ENERGY SAVING HEALTH CARE FACILITY — 012 建造节能型医疗设施
Why? — 012 为什么
- Enhanced Healing Environment — 012 强化康复环境
- Lower Construction Costs/Faster Payback — 013 降低建设成本 / 加速投资回报
- Reduced Operating Costs — 013 降低运营成本
- Reduced Greenhouse Gas Emmissions to Benefit the Community — 014 减少温室气体排放，造福社区

How? — 014 如何实现
- Benefits of Integrated Design — 014 综合设计的好处
- Features of Integrated Design — 016 综合设计的特征
- The Integrated Design Process — 017 综合设计的流程

RECOMMENDATIONS ON ENVELOPE — 020 建筑外壳设计建议
Opaque Envelope Components — 020 不传热建筑外壳组件
- Cool Roofs — 020 冷屋顶
- Roofs, Insulation Entirely above Deck — 022 楼顶上方的整体绝缘屋顶
- Walls – Mass — 022 大体块墙壁
- Walls – Steel Framed — 023 墙壁的钢架结构
- Below-grade Walls — 024 地下墙壁
- Floors – Mass — 024 大面积地面
- Floors-Metal Joist or Wood Joist/Wood Frame — 024 地面的金属托梁或木托梁 / 木结构
- Doors – Opaque, Swinging — 024 不传热转门
- Doors – Opaque, Roll-Up, or Sliding — 025 不传热卷帘门或拉门

Cautions — 025 注意事项
- Slab Edge Insulation — 025 楼板边缘绝缘
- Air Infiltration Control — 025 空气渗透的控制

Recommendations on Windows — 028 窗户设计建议
- Window Design Guidelines For Thermal Conditions — 028 窗户热状况设计指南
- Unwanted Solar Heat Gain is Most Effectively Controlled on the Outside of the Building — 028 在建筑外部可以有效地控制不必要的太阳辐射热增量
- Operable Versus Fixed Windows — 029 可控窗与固定窗
- Continuous Insulation to Avoid Thermal Breaks — 029 用连续绝缘来避免热隔断

Window Design Guidelines for Daylighting — 030 窗户自然采光设计指南
- Visual Transmittance — 030 可见光透射率
- Separating Views and Daylight — 031 分隔观景和采光
- Colour-Neutral Glazing — 032 中性色彩的玻璃装配
- Reflectivity of Glass — 032 玻璃的反射率
- Light-to-Solar-Gain Ratio — 032 光热比
- High Ceilings — 032 高天花板
- Light Shelves — 033 遮阳板

Recommendations on Lighting	034	照明设计建议
Electric Lighting	034	电气照明
Light-Colored Interior Finishes	034	浅色室内装饰
Linear Fluorescent Lamps and Ballasts	034	荧光灯管和镇流器
Compact Fluorescent	037	紧凑型荧光灯
Metal Halide	037	金属卤素灯
Exit Signs	038	出口标志
General Lighting Control Strategies	038	通用照明控制策略
Occupancy-Based Control	038	感应控制
Daylight Harvesting Control	039	日光收集控制
Electrical Lighting Design	040	灯光设计
Daylighting	040	日光照明
General Principles	040	总则
Consider Daylighting Early in the Design Process	042	在设计初期考虑日光照明
Use Daylighting Analysis Tools to Optimise Design	043	利用日光照明分析工具来优化设计
Space Types, Layout, and Daylight	044	空间类型、布局和日光
Building Orientation and Daylight	045	建筑朝向和日光
Building Shape and Daylight	046	建筑造型和日光
Window-to-Wall Ratio (WWR)	047	窗墙比
Sidelighting: Ceiling and Window Height	047	侧光照明：天花板和窗户高度
Sidelighting: Wall-to-Wall Windows	047	侧光照明：整墙窗户
NURSING HOMES		**疗养院**
Nursing Home Vivaldi	050	维瓦尔第疗养院
Asilo De Ancianos in Baños De Montemayor	062	蒙特马约尔老年疗养中心
Nursing Home Hainburg, Austria	070	奥地利海恩堡养老院
Pflegeheim St. Lambrecht	078	圣兰布雷希特临终关怀中心
Renovation and Enlargement of a Health Centre in Seregno (MB)	084	塞雷尼奥卫生中心改造扩建工程
Hanzeborg Care Centre	092	翰兹伯格护理中心
Pflegeheim Schorndorf	100	舍恩多夫疗养院
Specialised Clinic in Addiction Treatment Hoechsten, Bad Saulgau	108	巴德绍尔高－赫斯特恩戒瘾治疗中心
Boswijk Home	118	布什维基老年之家
FACILITIES FOR REHABILITATION		**康复设施**
Clinica San Pablo Chacarilla Physical Therapy Centre	128	圣巴勃罗查卡利亚物理治疗中心医务总部
Clinical Home Oegstgeest	136	乌赫斯特海斯特医疗中心
Mental Health Care: "High Care"	144	"高度关怀"心理健康中心
The Volgerlanden	152	国家心理健康护理中心
Extension of the Rehab Healthcare Facility	162	康复护理中心扩建工程
Wellness Centre Orhidelia	172	奥迪利亚疗养中心
Tokyo Bay Rehabilitation Hospital	180	东京湾康复医院
Children's Department and Work Therapy at the Institute for Rehabilitation of Republic of Slovenia, Ljubljana	192	斯洛文尼亚共和国卢布尔雅那康复研究所的儿科和工作治疗部
Rehab-Hotel Sonnenpark Rust	200	朝阳公园康复酒店
INDEX	206	**索引**

PREFACE Nursing Home
前言 疗养院

OVERVIEW

Nursing homes serve patients requiring preventive, therapeutic, and rehabilitative nursing care services for non-acute, long-term conditions. Specialised clinical and diagnostic services are obtained outside the nursing home. Most residents are frail and aged, but not bedridden, although often using canes, walkers or wheelchairs. Stays are relatively long, the majority for life. Nursing homes also care for a smaller percentage of convalescent patients of all ages. These patients are in long-term recovery from acute illnesses, but no longer require hospitalisation.

Nursing homes, or sections of them, are often classified into intermediate and skilled nursing units, definitions related to Medicare/Medicaid standards. Intermediate-care facilities have just enough nursing to qualify for Medicaid; skilled nursing facilities meet the more demanding medical standards to qualify for Medicare as well as Medicaid support. The cognitively impaired are frequently housed separately in Alzheimer Related Dementia (ASD) units. See Alzheimer's Foundation of America Excellence in Care Program[1].

Nursing homes present special design challenges in that for most residents the nursing home is not just a facility, but indeed their home. The reality is that in most cases the residents will live there for the rest of their lives and, moreover, rarely leave the premises at all. The nursing home becomes their entire world in a sense. The challenge is to design a nursing home that is sensitive and responsive to long-term human needs and well-being[2], both physical and emotional[3].

BUILDING ATTRIBUTES

A nursing home operates primarily in a patient-care mode rather than

概述

疗养院为那些需要接受慢性疾病预防、治疗和康复医护服务的病人而准备。专业的临床治疗和诊断服务由疗养院以外的医疗机构提供。在疗养院中居住的绝大多数是体弱和年长的人，虽然常常使用拐杖、助步车或轮椅，但是他们并没有卧床不起。他们在疗养院居住的时间相对较长，几乎占据余生的主要部分。疗养院也为一小部分处在康复期的、各年龄段的病人提供照顾服务。这些病人处在急性疾病的漫长恢复期中，但是不再需要住院治疗。

根据相关的医疗保险/医疗补助标准，疗养院，或者部分疗养院通常被归类为中级或熟练护理单位。中级护理设施仅有足够的、满足医疗救助资质要求的护理人员；熟练的护理设施需要满足更多医疗保险制度的医疗标准要求，同时符合医疗救助支持的要求。更多信息可参考阿兹黑梅尔基金美国优秀护理项目[1]。

对绝大多数住在疗养院里的人来说，疗养院不只是一个服务设施，更是他们的家，因此，疗养院是设计师们面临的特殊挑战。事实上，绝大多数居住者将在疗养院度过余生，此外，极少有提前离开的。疗养院在某种意义上成为他们的全部世界。设计者面临的挑战是设计出一个能对人类生理和情绪[3]的长期需求与幸福安宁[2]做出敏锐回应的疗养院。

建筑属性

疗养院的运行主要是对病人的护理，而不是治疗。因

a medical mode. Consequently, its more important attributes are those focusing on the general well-being of its residents rather than high-tech considerations. The principal attributes of a well designed nursing home are:

Homelike and Therapeutic Environment

Inherent in any institutional stay is the impact of environment on recovery, and the long-term stays typical of nursing home residents greatly increase this impact. The architect and interior designer must have a thorough understanding of the nursing home's mission and its patient profile. It is especially important that the design address aging and its accompanying physical and mental <u>disabilities</u>[4], including loss of visual acuity. To achieve the appropriate nursing home environment every effort should be made to:
• Give spaces a homelike, rather than institutional, size and scale with <u>natural light</u>[5] and views of the outdoors
• Create a warm reassuring environment by using a variety of familiar, non-reflective finishes and cheerful, varied colours and textures, keeping in mind that some colours are inappropriate and can disorient or agitate impaired residents
• Provide each resident a variety of spatial experiences, including access to a garden and the outdoors in general
• Promote traditional residential qualities of privacy, choice, control, and personalisation of one's immediate surroundings
• Alleviate possible disorientation of residents by providing differences between "residential neighbourhoods" of the nursing home, and by use of clocks, calendars, and other "reminders"
• Encourage resident autonomy by making their spaces easy to find, identify, and use
• Provide higher <u>lighting</u>[6] levels than typical for residential occupancies

此，其更重要的属性是关注居住者的整体健康，而不是高科技事项。一个设计良好的疗养院的主要特性如下。

像家一样的治疗环境

任何医疗规定的停留必然产生环境对康复的影响，而长期停留，特别是居住在疗养院的病人大大增强了环境对康复的影响力。建筑师和室内设计师必须对疗养院的使命及其病人的情况有一个彻底的了解。设计解决衰老和伴随衰老而来的生理和心理<u>残疾</u>[4]，包括视力减退，尤为重要。要实现适当的疗养院环境，应竭尽全力：
• 让空间像家一样温暖亲切，而不是刻板的医疗环境，有<u>自然光照明</u>[5]，可以看到户外的景色
• 通过多种常见的、非反射性的装饰和活泼多样的色彩与纹理营造一个温暖、安心的环境。但应谨记有些色彩并不适合使用，可能使患病的人产生迷惑或受到刺激
• 为每一个居住在疗养院的人提供各种各样的空间体验，一般包括进入花园、走向户外
• 发扬传统住宅在隐私、选择、控制和个人周围环境个性化方面的品质，将其融入疗养院的设计
• 通过疗养院"居住者邻里关系"之间的差异性，使用时钟、日历和其他"提示"，尽可能减轻疗养院居住者对方向的迷惑感
• 通过使他们的空间易于发现、辨识和使用，鼓励疗养院的居住者们实现生活自治
• 提供高于一般住宅的<u>照明</u>[6]水平

PREFACE Nursing Home 前言 疗养院

Efficiency and Cost-Effectiveness[7]
The nursing home design should:
• Promote staff efficiency by minimising distance of necessary travel between frequently used spaces
• Allow easy visual supervision of patients by minimal staff
• Make efficient use of space[8] by locating support spaces so they may be shared by adjacent functional areas, and by making prudent use of multi- purpose spaces

Cleanliness and Sanitation
An odour-free environment is a very high priority in nursing homes, since many residents are occasionally incontinent, and the pervasive odours can give an impression of uncleanliness and poor operation to family and visitors. In addition to operational practices and careful choice of furniture, facility design can help odour control by:
• Adequate and highly visible toilet rooms in key locations near spaces where residents congregate
• The use of appropriate, durable finishes for each space used by residents
• Proper detailing of such features as doorframes, casework, and finish transitions to avoid dirt-catching and hard-to-clean crevices and joints
• Adequate and appropriately located housekeeping spaces
• Effective ventilation, which may need to exceed nominal design levels
• Incorporating O&M practices[9] that stress indoor environmental quality (IEQ[10])

Attention to Way-finding
A consistent and well thought out system of way-finding helps to maintain the residents' dignity and avoid their disorientation. It should:
• Use multiple cues from building elements, colours, texture, patterns,

功效与成本效率[7]
疗养院的设计应：
・通过尽量缩短常用空间之间的距离，提升员工的工作效率
・便于员工对病人进行监护
・通过辅助型空间布局提升空间[8]的使用效率，相邻的功能区可对其进行共享，谨慎利用多功能空间

整洁与卫生
对疗养院来说，无气味的环境是极其重要的，因为许多居住者偶尔会失禁，充斥在房间中的气味会给人不洁的印象，来访者也会认为疗养院经营不善。除了运营实践，谨慎选择家具外，设施环境设计可以通过以下方式进行气味控制：
・在居住者聚集的空间附近的关键位置设置足够的、易于找到的卫生间
・居住者使用的每个空间都选用了合适的并且耐用的装饰材料
・像门框、橱柜这类表层的细节设计应恰当，避免容易沾染污垢、不易清洁的裂缝和接合处，进行修饰过渡
・清洁内务管理空间应足够，并且位置设计恰当
・有效的通风，可以高于理论设计标准
・结合操作与维护实践[9]，增强室内环境质量[10]

关注环境导示
协调一致，并经深思熟虑设计而成的环境导示系统有助于维护疗养院居住者的尊严，避免他们迷失方向，这一导示系统应：

and artworks, as well as signages, to help residents understand where they are, what their destination is, and how to get there and back.
• Identify frequently used destination spaces by architectural features and landmarks which can be seen from a distance, as well as symbols, signage, art, and elements such as fish tanks, birdcages, or greenery
• Avoid prominent locations and high visibility of doors to spaces which patients should not enter
• Use simple lettering and clear contrasts in signage (See VA Signage Manual[11])
• Clearly identify only those rooms that residents frequent

Accessibility

Many residents may be ambulatory to varying degrees, but will require the assistance of canes, crutches, walkers, or wheelchairs. To accommodate these residents, all spaces used by them, both inside and out, should:
• Comply with the requirements of the Americans with Disabilities (ADA[12]) and, if federally funded or owned, the GSA's ABA Accessibility Standards[13]
• Be designed so that all spaces, furnishings, and equipment, including storage units and operable windows, are easily usable by residents in wheelchairs
• Be equipped with grab bars in all appropriate locations
• Be free of tripping hazards
• Be located on one floor if feasible, preferably at grade. If residents' bedrooms must be located on more than one floor, then dining space must be apportioned among those floors, not centralised

Security and Safety

Design to address security and safety concerns of nursing homes includes:
• Use of non-reflective and non-slip floors to avoid falls

・利用建筑元素、色彩、纹理、图案和艺术品以及符号等多种提示，帮助疗养院的居住者了解其所在地，知道自己要去什么地方，如何到达并且返回
・通过可以远距离看到的建筑特征和地标物以及标识、符号、艺术品和诸如鱼缸、鸟笼或温室等元素对经常被使用的空间进行辨识、确认
・那些病人不能进入的空间，设计者应避免突出其位置、设置明显的门和通道
・标识采用简单的字体和鲜明的对比（见《美国退伍老兵标识设计手册》[11]）
・仅对疗养院居住者经常去的房间进行明显的标识

可抵达性

很多居住在疗养院里的人的行走能力不同，但是会需要使用手杖、拐杖、助步车或轮椅。这些人使用的所有空间，无论室内和室外都应：
・符合《美国残疾人法案》（ADA）[12]，如果疗养院属联邦出资或所有，还应符合美国一般服务管理建筑障碍物法案对可及性的规定[13]
・经精心设计，这样所有的空间、陈设和设施，包括储藏间和可操作窗体均便于坐轮椅的人使用
・在所有适当的位置加装扶手杆
・处在同一楼层（如果可行的话），最好在同一水平面上。如果疗养院里的卧室布局超出一层楼，那么就餐空间必须分布在这些楼层中间，而不是集中在一处

安保与安全

设计疗养院时涉及的安保与安全问题包括：
・使用非反射性的、防滑地板材料，防止滑倒

- Control of access to hazardous spaces
- Control of exits to avoid residents leaving and becoming lost or injured
- Provision of secure spaces to safeguard facility supplies and personal property of residents and staff

Aesthetics

Aesthetics is closely related to creating a therapeutic homelike environment. It is also a major factor in a nursing home's public image and is thus an important marketing tool for both residents' families and staff. Aesthetic considerations include:
- Increased use of natural light[14], natural materials, and textures
- Use of artwork
- Attention to proportions, colours, scales, and details
- Bright, open, generously scaled public and congregate spaces
- Homelike and intimate scale in resident rooms and offices
- Appropriate residential exterior appearance, not hospital-like
- Exterior compatibility with surroundings

Sustainability

Nursing Home facilities are public buildings that may have a significant impact on the environment and economy of the surrounding community. As facilities built for "caring", it is appropriate that this caring approach extend to the larger world as well, and that they be built and operated "sustainably".

Section 1.2 of VA's HVAC Design Manual is a good example of health care facility energy conservation standards that meet EPAct 2005[15] and Executive Order 13423[16] requirements. The Energy Independence and Security Act of 2007 (EISA)[17] provides additional requirements for energy

· 限制危险空间的可及性
· 管理出口，防止疗养院居民自行离开而失踪或受伤
· 为安保设施提供安全的空间，保护疗养院的居住者和工作人员私人财产安全

美学

美学与打造一个具有亲和力的治疗空间密切相关。美学也是疗养院公共形象的一个主要因素，因此对病人家属与员工来说还是一个重要的营销手段。疗养院的美学设计需要考虑的事项包括：
· 提升自然光[14]、天然材料和质地的利用率
· 使用艺术作品
· 注意分布、色彩、比例和细节
· 明亮、开阔、大开间的公用与聚集空间
· 病人房间和办公室空间比例的家居感和亲密感
· 适当的建筑外观，不要像医院一样
· 外观与周围环境的兼容性

可持续性

疗养院属于公共建筑，可能对环境和周围社区经济产生极大的影响。因为这类设施的建造是出于"人道"，所以最好还是将人文关怀的做法延伸到更大的世界，而且应当使它们的建造和运营具有可持续性。

《美国退役老兵事务管理机构关于采暖通风与空调设计手册》第2.1部分就是医疗保健设施能源保护标准的一个很好的范例，它符合2005能源法案[15]以及美国总统第13423号行政命令[16]的要求。《2007年能源独立与安全法案》[17]

Note (for more information please visit)（注：更多信息请访问）：
1. http://www.excellenceincare.org/
2. http://www.wbdg.org/design/promote_health.php
3. http://www.wbdg.org/resources/psychspace_value.php?r=nursing_home
4. http://dsc.ucsf.edu/main.php
5. http://www.wbdg.org/resources/daylighting.php?r=nursing_home
6. http://www.wbdg.org/resources/efficientlighting.php?r=nursing_home
7. http://www.wbdg.org/design/cost_effective.php
8. http://www.wbdg.org/design/spacetypes.php
9. http://www.wbdg.org/resources/sustainableom.php?r=nursing_home
10. http://www.wbdg.org/design/ieq.php
11. http://www.wbdg.org/ccb/browse_cat.php?o=34&c=22
12. http://www.ada.gov/
13. http://www.access-board.gov/ada-aba/aba-standards-gsa.cfm
14. http://www.wbdg.org/resources/daylighting.php?r=nursing_home
15. http://www.gpo.gov/fdsys/pkg/BILLS-109hr6enr/pdf/BILLS-109hr6enr.pdf
16. http://www.wbdg.org/ccb/browse_doc.php?d=7707
17. http://www.gpo.gov/fdsys/pkg/BILLS-110hr6enr/pdf/BILLS-110hr6enr.pdf
18. www.usgbc.org/DisplayPage.aspx?CMSPageID=1765
19. http://www.hipaa.org/

conservation. Also see LEED's (Leadership in Energy and Environmental Design) USGBC LEED for Healthcare[18].

Related Issues
The HIPAA[19] (Health Insurance Portability and Accessibility Act of 1996) regulations address security and privacy of "protected health information" (PHI). These regulations put emphasis on acoustic and visual privacy, and may affect location and layout of workstations that handle medical records and other patient information, paper and electronic, as well as patient accommodations."

Emerging Issues
There is a growing recognition of the need for dementia day care. This can often be effectively provided within or adjoining an inpatient nursing facility.

There is a need for better non-medical residential facilities for the frail but independent elderly. Managed care programs for the aged are being developed to prevent, or at least postpone, institutionalisation.

by Robert F. Carr
NIKA Technologies, Inc. for VA Office of
Construction & Facility Management (CFM)
Revised by the WBDG Health Care Subcommittee

提供了关于能源保护的其他要求。也可参考LEED卫生保健认证（领先能源与环境设计建筑评级体系）[18]。

相关问题
《1996美国医治保险携带和责任法案》（HIPAA）[19] 规定"保护健康信息安全"（PHI）。这些规定着重于声音和视觉隐私保护，因此可能影响那些处理病历和其他纸质和电子版病人信息、安排病人食宿的工作站的选址和布局。

暴露的问题
对痴呆病人日常护理的需求日益增长已有目共睹。这通常可在住院病人护理设施内或临近的设施中得到有效解决。

对身体虚弱需要照顾的老人来说，需要的是一个居住设施而非医疗设施。养老托管护理项目正在制定，以起到预防，至少是延迟的作用，使其制度化。

罗伯特·F·卡尔
NIKA技术有限公司，为美国退伍老兵建筑与设施管理办公室撰写，由卫生保健整体建筑设计指导委员会修订

To Build Up Energy Saving Health Care Facility
建造节能型医疗设施

1. WHY?

1.1 Enhanced Healing Environment

A healthcare facility that includes favourable light, sound, and temperature provides a better experience for patients and their families by enhancing comfort and control while reducing stress and anxiety.

Access to views and daylighting, which uses the sun to produce high-quality, glarefree light in a space, has been found to favorably affect both patient outcomes and staff productivity. In a recent report by R.S. Ulrich, "How Design Impacts Wellness,"[1] it was found that a patient room providing good outdoor views and daylighting can increase patient well-being and create a psychological state resulting in reduced stress and anxiety, lower blood pressure, improved post-operative recovery, reduced need for pain medication, and shorter hospital stays. Daylighting will also significantly reduce ambient electric light energy consumption; lighting power savings during daylight hours in controlled spaces can be as high as 87%.

Related research also shows that the ability to control their personal environment, including bedside control of lighting and window shades, can improve patients' psychological outlooks, rates of healing, and quality of stay.

According to the American Society of Healthcare Engineers (ASHE), the health of patients, staff, and visitors can be profoundly affected by the quality of the indoor air. A recent study completed by the Lawrence Berkeley National Laboratory (LBNL) reported that improvements to indoor environment could reduce healthcare costs and work losses from communicable respiratory disease by 9% to 20%. Advanced, energy-efficient heating and cooling systems can also create cleaner, healthier indoor environments that reduce the threat of infection for both patients and staff.

Advanced energy-efficient systems can also be much quieter than previous technology.

This produces quieter, more comfortable and more productive spaces. This all translates to better patient outcomes, shorter patient stays, reduced sick-days for healthcare staff, and lower overall costs.

1. 为什么

1.1 强化康复环境

在医疗设施里良好的光线、声音和温度能够增加舒适度、加强管理、减少压力和焦虑，从而为患者及其家属提供更好的体验。

研究表明，良好的视野和自然采光（利用太阳所营造的高质量、不刺眼的光线）对患者康复和医务人员的工作效率都有积极的作用。R·S·乌利齐最近的一份报告《设计如何影响健康》[1]显示，拥有良好室外视野的自然采光的病房能够促进患者痊愈，其所营造的心理环境能够减少压力和焦虑、降低血压、促进术后康复、减少止痛药需求、缩短住院时间。自然采光还能大幅度减少电灯的能源消耗；白天的照明能源最多可节约87%。

相关研究还表明，对身边环境（如在床侧控制灯光和窗帘）的控制能够提升患者的心理感受、促进痊愈并且提高住院质量。

美国健康工程师协会称，室内空气质量能够显著地影响患者、医务人员以及访客的健康状况。劳伦斯·伯克利国家实验室最近有研究表明，室内环境质量的提升可以将呼吸道传染疾病的医疗成本和误工损失降低9%到20%。先进的高能效采暖和制冷系统同样也能营造出更清洁、更健康的室内环境，从而减低患者和医务人员的感染风险。

改进的节能系统比早期的技术更静音。

这些都能营造更安静、更舒适、更高效的空间，从而为患者提供更好的治疗效果、缩短其住院时间、减少医务人员的患病几率、减少整体成本。

1.2 Lower Construction Costs/Faster Payback

Thoughtfully designed, energy-efficient hospitals can cost less to build than typical hospitals. For example, optimising the envelope to match the climate can substantially reduce the size of the mechanical systems. A hospital with strategically designed glazing will have lower mechanical costs than the one without, and will cost less to build. In general, an energy-efficient hospital
- requires less heating,
- costs less to maintain,
- has less expensive installation,
- requires fewer lighting fixtures due to more efficient lighting,
- allows for downsized heating systems due to better insulation and windows, and
- allows for downsized cooling systems with a properly designed daylighting system and a better envelope.

Some strategies may cost more up front, but the energy they save means they often pay for themselves within a few years.

1.3 Reduced Operating Costs

According to the most recent Commercial Buildings Energy Consumption Survey (CBECS) conducted by the Energy Information Administration (EIA), the average hospital in North America consumes nearly 250% more energy than the average commercial building. By using energy efficiently and lowering a hospital's energy bills, hundreds of thousands of dollars can be redirected each year into upgrading existing facilities, caregivers' salaries, and investing in the latest technology in medical equipment.

Strategic up-front investments in energy efficiency provide significant long-term savings. In the financial performance of the hospital, every dollar saved in energy and operating costs is equal to generating $20 growth in new top line revenues.

In an average hospital, lighting uses a large portion of the overall energy budget. Therefore the design should include an energy-efficient lighting design and efficient lighting fixtures but also should evaluate opportunities for dimming controls and multilevel switching systems. In the many areas where the design team brings quality daylight into the space, lighting

1.2 降低建设成本／加速投资回报

通过巧妙的设计，节能型医院的建设可以比普通医院更加经济。例如，优化建筑外壳使其适应环境能够大幅缩小机械系统的规模。拥有良好玻璃装配的医院的机械成本要低于没有经过设计的医院，同时它的建造成本也低得多。总体来讲，节能型医院拥有以下优点：
–采暖需求低
–维护成本低
–设施安装成本低
–由于采用了高效照明，所需的灯具数量较少
–良好的隔热和窗户设计能够缩减采暖系统规模
–设计良好的照明系统和优良的建筑外壳能够缩减制冷系统的规模
一些策略可能前期成本较高，但它们所节约的能源会在几年内抵消成本。

1.3 降低运营成本

根据美国能源信息管理局最新的商业建筑能源消耗调查，北美医院的能耗比商业建筑能耗平均值高250%。通过采用高效能源设备，缩减医院能耗账单，医院每年能够节省下数万美元用于升级现有设施、支付看护者的工资以及投入到最先进的医疗设备中。

在能源效率方面的策略性前期投资能够实现长期的节约。在医院的财政业绩中，在能源和运营成本中节省下来的1美元相当于顶端收益增长20美元。

在医院中，照明占据了能源开支的一大部分。因此，在对医院进行设计时，应当包括节能照明设计和对高效灯具的合理选择，同时，还应当评估调光控制和多级开关系统应用的可能性。在许多采用高品质自然采光的空间里，照明控制可以用于调节电灯的输出功

controls can be used to regulate the output of electric lights to optimize the quality of the visual environment while saving significant amounts of energy.

The smart use of a site's climatic resources and more efficient envelope design are keys to reducing a building's overall energy requirements. Efficient equipment and energy management programmes then help meet those requirements more cost-effectively.

Because of growing water demand and shrinking aquifers, the price of water is escalating in many areas. Saving water can thus save money but also can generally save energy.

Lower operating costs mean less fluctuation in budgets because of price instabilities of energy. Purchasing energy efficiency is buying into energy futures at a known fixed cost.

1.4 Reduced Greenhouse Gas Emmissions to Benefit the Community
According to some estimates, buildings are responsible for nearly 40% of all carbon dioxide emissions annually in the United States. Carbon dioxide, which is produced when fossil fuel is burned, is the primary contributor to greenhouse gas emissions.

Healthcare facilities can be a part of the solution when they reduce their consumption of fossil fuels for heating, cooling, and electricity. The local community, patients, and staff will appreciate such forward-thinking leadership.

2. HOW?
Integrated Process for Achieving Energy Savings

2.1 Benefits of Integrated Design
The primary mission of healthcare is healing, and of hospitals and healthcare facilities it is first to do no harm. For healthcare facilities, the emphasis on achieving energy savings is still relatively new, as compared to other building types. Healthcare design is strongly impacted by building

率，从而优化视觉环境的质量、大幅度减少室内的能源消耗。

巧妙利用场地的气候资源和高效的建筑外壳是设计师减少建筑整体能源需求的关键。高效的设备和能源管理项目能够实现这些需求。

由于人们不断增长的用水需求和逐渐萎缩的蓄水层，水价在许多地区不断攀升。节水就是省钱，同时也是在节约能源。

降低运营成本意味着由能源价格不稳定性所引起的预算波动大幅减少。投资能源效率相当于以固定的价格购买能源的未来。

1.4 减少温室气体排放，造福社区
根据一些预测数据，在美国，建筑所排放的二氧化碳接近二氧化碳排放总量的40%。化石能源燃烧所产生的二氧化碳是温室气体的主要组成部分。

如果医疗设施减少它们在采暖、制冷和电力方面的化石能源消耗，这一问题将得到一定程度的解决。当地社区、患者和医务人员都会感激这种有远见的行动。

2. 如何实现
实现能源节约的综合流程。

2.1 综合设计的好处
医疗保健的首要任务是治疗，对医院和医疗设施来说，首先便要保持环境的无害化。与其他设施相比，医疗设施的设计对节能的考虑相对较新。医疗设计必须严格遵守建筑规范、许可要求以及医学规划，这些都为实现节

codes, licensing requirements, and medical planning, making it more challenging to justify energy efficiency goals. However, these goals can gain additional weight when supported by their ancillary benefits such as improved indoor environment, improved staff satisfaction, improved patient outcomes, and other benefits inherent in many energy reduction strategies.

Using integrated design will allow better ability to achieve multiple design goals without adversely impacting first cost. For example, a 2006 study by Matthiessen and Morris concluded that many green building strategies can be implemented with minimal or no additional cost, while some can reduce first cost through improved design or reduced complexity of design.

A largely untapped benefit arising from energy efficiency is reduced building size or more usable space. A typical building of this type might include 16 ft floor-to-floor height with 9 ft ceilings. This volume, together with spaces used to house mechanical and electrical equipment, means that more than 40% of the building volume is building space dedicated to ductwork and equipment. The net usable space is barely more than half of the building volume. Integrated design can achieve less overhead space and more net usable space.

Much of the HVAC energy consumption in a healthcare building is driven by the various code requirements, operating schedules, and medical equipment. Even so, optimising the shape of the building, the envelope, and architectural planning can minimise additional energy loss. Moreover, the way a room is defined by the architecture can have a dramatic impact on the code-required HVAC parameters for a particular space. The integrated design process targets the energy demand side by lowering envelope and interior building loads of the building, optimising site layout and building shape and orientation, and increasing envelope thermal efficiency. This reduces the demands for the subsystems such as HVAC, lighting, plumbing, and power. Integration allows the "rightsizing" of building systems and components, which reduces first and life-cycle costs.

能目标提出了挑战。然而，这些目标的辅助价值可以实现额外的获利，例如良好的室内环境、员工满意度的提升、良好的治疗效果以及其他节能策略所带来的好处。

综合设计能够更好地实现多重设计目标，同时又不会影响初始成本。例如，马西森和莫里斯在2006年的研究指出：许多绿色建筑策略的实施只需要极少、乃至不需要附加成本；一些策略还能通过改进设计和简化设计来降低初始成本。

节能设计所带来的诸多好处之一，就是能够缩减建筑规模或者增加可用空间。此类型的建筑通常层高16英尺（约4.88米），天花板高9英尺（约2.74米）。吊顶空间与机械和电力设备所用的空间一起，意味着管道设施和相关设备占用了40%以上的建筑空间。可用空间仅为建筑空间的一半。综合设计能够减少头顶空间，提供更多的可用空间。

医疗建筑内空调系统的能耗大都由各种建筑规则要求、工作时间安排和医疗设备决定。尽管如此，优化建筑造型、外壳以及建筑规划还是能将额外的能耗最小化。此外，建筑内房间的设计对特定空间的空调系统参数有着强烈的影响。综合设计流程通过降低建筑的外壳和室内负载、优化场地布局、建筑造型和朝向、增加建筑外壳热能效率来实现节能要求。这将减少空调、照明、管道、电力等次级系统的能源要求。综合设计实现了建筑系统和组件的规模优化，减少了初始成本和寿命周期成本。

2.2 Features of Integrated Design

A successful integrated design approach provides the best energy performance at the least cost and is characterised as follows.

It is goal driven. In a goal-setting session early in the design process, strategies are identified to meet energy-efficiency goals in relation to the owner's mission. Goals must be quantifiable and measurable. Defining energy performance as a project goal from the beginning of the project is necessary to ensure that the energy performance is equally prioritised in the design and prevails throughout the project. By including medical equipment planners, the client's user groups, and engineering and facility departments in this session, each group's interests can be aligned with the project goals and successful implementation into the project can be ensured. Selecting medical equipment planners who consider energy consumption can have a substantial impact on facility energy use.

It is resourceful. Integrated design begins with site assessment and site layout studies to optimize orientation and fenestration for the best light to solar gain ratio. Layout and orientation are opportunities to obtain free energy resources. As an example, some of the artificial lighting can be replaced by daylight through the windows. Passive solar strategies such as direct gain can reduce mechanical heating energy consumption in colder climates, and passive shading can reduce cooling energy consumption. On the supply side, geothermal and solar energy (domestic hot water, heating water, and/or photovoltaic systems) reduce the amount of energy required from fossil fuels. Optimal building orientation, form, and layout achieve substantial energy savings.

It is multidisciplinary. Integrative design is a radically different process to the conventional approach used for project design and delivery. The traditional practice relies on isolated specialists, each optimising their own systems, and can result in component and equipment sizing by rule of thumb and vastly oversizing systems. Instead, integrated design involves the owner, designers, technical consultants, construction manager (CM), general contractor (GC), CxA, facility staff, and end users in all phases of the project working to optimize the whole design. The process requires cross-disciplinary design and validation at all phases of the process.

2.2 综合设计的特征

成功的综合设计方案能够以最低的成本提供最佳的能源性能，它拥有以下几个特征。

以目标为导向。在设计初期的目标设定阶段，综合设计就已经依照业主需求确定了能效目标。目标必须是量化、可测量的。在项目初期确定以能源性能为目标能够有效保证其在整个设计中的重要地位。通过将医疗设备规划人员、委托人的用户群、工程师和设备部门集中在这一阶段，每个群体的利益都与项目目标联系起来，保证了项目目标的实现。能够对能源消耗进行考量的医疗设备规划人员将对医疗设施的能源有效利用有所帮助。

方式丰富多样。综合设计从场地评估和场地布局研究开始，以便优化建筑朝向和门窗布局，有利于最佳光热比的实现。布局和朝向设计为获取免费能源提供了机会。例如，一些人工照明可以被自然光所取代。直射等被动式太阳能策略可以减少寒冷天气中的机械采暖能耗，而被动式遮阳则能减少制冷能耗。在供给方面，地热和太阳能（家用热水、水加热、光伏系统）可以较少化石燃料的消耗。优化建筑朝向、造型和布局能够节约大量能源。

具有跨学科性。综合设计与传统项目设计和交付使用有着巨大的区别。传统设计实践依靠独立的专家，他们各自为政，可能会凭经验对元件和设备进行设计，从而形成规模过大的系统。相反，综合设计让业主、设计师、技术顾问、施工经理、总承包商、验收人、机构员工以及终端用户参与到项目设计的各个流程之中，优化了整个设计。整个过程要求进行跨学科设计和验收。验收方可以是医疗机构的工作人员、设计公司的独立工作人员或是外部顾问，他是反复设计过程中的重要组成部分。

The CxA, who may be a member of the healthcare facility staff, an independent staff member from the design firm, or an outside consultant, is an integral part of this iterative process. He or she validates that the design documents meet the energy savings goals; that the building is constructed as designed; and that the staff knows how to use, operate, and maintain the building to achieve the energy savings goals.

It is iterative. A goal-setting session is just the start. As the design concept takes shape, strategies need to be tested to determine if the results meet the desired energy performance targets and whether maintenance requirements need optimising and lifecycle costs need reducing. Even when a team follows prescriptive energy-saving strategies such as those in this document, energy modelling can provide further refinement and optimised performance. It is imperative to make energy performance a standing agenda item during design reviews to discuss trade-off opportunities at the system level.

2.3 The Integrated Design Process

The following description delineates an integrated design process for tracking and achieving energy savings in new small healthcare facilities, outlining the involvement of owners, architects, consultants, and builders who decide to augment and improve their practices to include energy efficiency at each stage of the development process from project conception through building operation. In the following tables and throughout this Guide, the expression consultants is used as an abbreviated term defined as including special consultants from mechanical, electrical, and all other engineering disciplines.

2.3.1　1. Pre-Design (PD) (or Planning and Programming) Phase

Adopting measurable energy goals at the beginning of the project will guide the team and provide a benchmark during the project's life. General strategies that relate to these goals will be identified at this phase as part of the goal discussions. Strategies will be further refined and confirmed during the design phase.

Analyse the site and programme to identify the largest savings potentials and focus attention on these first. Priorities are likely to vary significantly

他将确认设计文件是否符合节能目标、建筑是否按照设计进行施工，保证员工了解如何使用、运行和维护建筑，以实现节能目标。

具有反复性。目标设定环节仅仅是一个开始。随着设计概念的成形，项目需要对策略进行测试，验证期是否与预期的能效目标相符、维护需求是否需要优化、是否需要减少寿命周期成本等。即使团队严格依照文件中指定的能源策略进行实施，能源建模也能够提供进一步的优化和改进意见。有必要将能源性能作为一个长期的议程项目，在设计评估中进行讨论，从系统的层面上进行权衡取舍。

2.3 综合设计的流程

以下文字描述了综合设计的流程，包括在新建小型医疗设施中追踪和实现节能目标，概述业主、建筑师、工程顾问、建筑商从项目概念到建筑运营过程中的探讨及改进过程。在下文中，"顾问"包括机械、电气以及所有其他工程领域的专业顾问。

2.3.1 预设计阶段（规划阶段）

在项目初期采用可衡量的能源目标将为团队提供指导，为项目提供终生的设计标杆。这一阶段将在目标讨论中明确与这些目标相关的总体策略。在设计阶段，这些策略将得到进一步的改进和确定。

团队首要考虑的就是评估场地和项目规划，以实现最大的节约潜能。在不同的气候带、同一气候带的不同

from one climate zone to another and may vary between small healthcare facilities in the same climate zone. Site conditions can significantly affect energy performance. For example, differences in building application, climate, and even orientation will affect the selection of various energy goals and strategies.

Because of the high air change rates and humidity control required in many of the space types found in healthcare facilities, the constant-volume reheat HVAC systems that have traditionally been used in these facilities use a lot of energy for reheat. The baseline energy modeling for this Guide shows that reheat represents over 20% of the total energy use of the building, and this occurs in all climate zones.

2.3.2 Design Phase

In the design phases (schematic design and design development), the project team develops energy efficiency strategies and tests them for compliance with project goals before incorporating them into building drawings and specifications. Systems are optimised in a systemic way and selected based on their aggregate performance rather than evaluating component by component. Strategy choices should be prioritized according to their overall efficacy in energy consumption reduction. Selecting and prioritising energy conservation measures need to include consideration of the impact on the owner's facility operating staff. Some strategies would require additional staff with increased capabilities.

- Select energy-efficient mechanical systems
- Reduce thermal building loads
- Optimise on-site energy resources
- Size systems to comply with reduced loads
- Incorporate efficient mechanical equipment and lighting

At each point, the decisions should take into account priorities and systems decisions. For example, cooling system sizing should take into account daylighting measures, glazing sizes, and building orientation.

The CxA reviews the design to verify that the project goals are being met. The CxA should also verify that the assumptions for HVAC load

医疗机构，优先顺序都不尽相同。场地条件能够显著影响能源性能。例如，建筑用途、气候乃至朝向的不同都会影响能源目标和策略的选择。

由于医疗设施的许多空间都需要较高的空气交换率和严格的湿度控制，传统医疗设施中所使用的恒容再热空调系统会采用大量能源进行再热。本文中的能源建模基准线显示：在所有气候带，再热占据了建筑总能耗的20%以上。

2.3.2 设计阶段

在设计阶段（方案设计和设计开发），项目团队将开发节能策略，并在将其运用到建筑图纸和规范之前检验它们是否与项目目标一致。各个系统将以系统的方式进行优化，以其总体表现进行选择，而非单独考虑各个元素。策略的选择将根据它们在缩减能源消耗上的总体绩效确定优先权。节能策略的选择和排序还需要考虑其对设施管理人员的影响。一些策略可能会要求额外的人员来实现。

* 选择节能机械系统
* 减少建筑热负载
* 优化现场能源资源
* 根据较少的负载调整系统规模
* 引用高效机械设备和照明

每一项决定都应当考虑优先权和系统决策。例如，制冷系统的规模设计应当考虑日光照明措施、玻璃装配尺寸和建筑朝向。

验收人将对设计进行复核，确定其是否与项目目标相符。验收人还应当检验预期空调荷载及其他模型预期结

Note: 1. ASHRAE, 2007 ASHRAE Handbook – HVAC Applications (Atlanta: American Society of Heating, Refrigerating and Air-Conditioning Engineers, Inc., 2007).

Excerpted from Advanced Energy Design Guide for Small Hospitals and Healthcare Facilities © 2009 American Society of Heating, Refrigerating and Air-Conditioning Engineers, Inc. (www.ashrae.org)

注：1. 美国采暖、制冷与空调工程师学会，2007美国采暖、制冷与空调工程师学会手册——空调系统应用（亚特兰大：美国采暖、制冷与空调工程师学会，2007）

摘录自小型医院与卫生保健设施先进节能设计指南©美国采暖、制冷与空调工程师学会(www.ashrae.org)

calculations and other modelling assumptions are based on actual design parameters rather than on rule of thumb.

2.3.3 Construction Phase

Even the best design will not operate successfully and yield the expected energy savings if the construction drawings and specifications are not correctly executed.

2.3.4. Acceptance, Occupancy, and Operation Phase

Occupancy is a critical time in the process, and is often neglected by the project teams.

Energy savings are difficult to attain if the medical, engineering, and O&M staff do not know how to use, operate, and maintain the building. The CxA should ensure timely submittals of the O&M manuals through specifications and regular reminders at construction meetings and ensure adequate and timely training of all facility and medical staff. A performance review should be conducted during the first year of building operation. The building operator should discuss any systems that are not performing as expected with the design and construction team so they can be resolved during the warranty period.

Over time, the building's energy use, changes in operating hours, and any addition of energy-consuming equipment should be documented by the owner's facilities staff. This information can be used to determine how well the building is performing and for taking lessons back to the design table for future projects. Performance evaluations should take place on a schedule specified in a maintenance manual provided to the owner as part of final project acceptance. Ongoing training of facility staff, including medical staff, administrators, and instructional staff should be provided to such address changes and to address staff turnover.

果是以实际设计参数为基础而非凭经验想象得来的。

2.3.3 施工阶段

如果施工图纸和相关标准没有得到正确的实施，再好的设计也不会成功，而预期的节能目标也将无法实现。

2.3.4 验收、投入使用和运行阶段

投入使用是一个关键的流程，常常被项目团队所忽略。

如果医务人员、工程人员和运行维护人员不知道如何使用、运营和维护建筑，节能目标就很难实现。验收人应当保证运行维护手册和施工会议上所提出的注意事项能够及时提交，保证所有物业和医务人员得到及时、充分的培训。在建筑运行的第一年应当进行一次性能评估。建筑经营者应当对没有按设计和施工团队预期运行的系统进行讨论，以便在保修期内解决问题。

业主的工作人员应当将建筑能耗、运行时间的变化以及任何耗能设备的添加记录在案。这些信息可以用来评估建筑的运行情况，并且能对未来项目的设计提供经验。项目验收时，项目团队应向业主提供一本维护手册，建筑的性能评估应当在这个手册内进行记录。

Recommendations On Envelope
建筑外壳设计建议

1. OPAQUE ENVELOPE COMPONENTS

1.1 Cool Roofs
To be considered a cool roof, a Solar Reflectance Index (SRI) of 78 or higher is recommended.

A high reflectance keeps much of the sun's energy from being absorbed while a high thermal emissivity surface radiates away any solar energy that is absorbed, allowing the roof to cool more rapidly. Cool roofs are typically white and have a smooth surface. Commercial roof products that qualify as cool roofs fall into three categories: single-ply, liquid-applied, and metal panels. Examples are presented in Table 1.

The solar reflectance and thermal emissivity property values represent initial conditions as determined by a laboratory accredited by the Cool Roof Rating Council (CRRC). An SRI can be determined by the following equations:

$$SRI = 123.97 - 141.35(x) + 9.655(x^2)$$

Where

$$x = \frac{20.797 \times \alpha - 0.603 \times \epsilon}{9.5205 \times \epsilon + 12.0}$$

and

α = solar absorptance = 1 - solar reflectance
ϵ = thermal emissivity
These equations were derived from ASTM E1980 assuming a medium wind speed.
Note that cool roofs are not a substitute for the appropriate amount of insulation.

1. 不传热建筑外壳组件

1.1 冷屋顶
冷屋顶的日光反射率应当为78或78以上。

高反射率可以阻止太阳能的吸收,高性能热辐射表面还能将其所吸收的太阳能辐射出去,让屋顶快速冷却。冷屋顶基本为白色,表面光滑。冷屋顶类的商业屋顶产品分为三种:单板层、液体附着板和金属板。举例详见表1。

日光反射率和热辐射指标是冷屋顶评级委员会实验室认可的初始参数。日光反射率由以下公式得出:

$$日光反射指数 = 123.97 - 141.35(x) + 9.655(x^2)$$

其中

$$x = \frac{20.797 \times \alpha - 0.603 \times \epsilon}{9.5205 \times \epsilon + 12.0}$$

并且

α = 日光吸收比 = 1 - 日光反射系数
ϵ = 热发射率
这些公式源于ASTM标准E1980,假设风速中等;注意冷却式屋顶不能替代适当的隔热处理。

Table 1 – Examples of Cool Roofs
表1-冷却式屋顶示例

Category 类别	Product 产品	Reflectance 反射系数	Emissivity 发射率	SRI 日光反射指数
Single-ply 单层	White polyvinyl chloride (PVC) 白色聚氯乙烯（PVC）	0.86	0.86	107
	White chlorinated polyethylene (CPE) 白色氯化聚乙烯（CPE）	0.86	0.88	108
	White chlorosulfonated polyethylene (CPSE) 白色氯磺化聚乙烯（CPSE）	0.85	0.87	106
	White thermoplastic polyolefin (TSO) 白色热塑性聚烯烃（TSO）	0.77	0.87	95
Liguid-applied 液体附着	White elastomeric, polyurethane, acrylic coating 白色弹性、聚氨酯、丙烯酸涂料	0.71	0.86	86
	White paint (on metal or concrete) 白色涂料（用于金属或混凝土）	0.71	0.85	86
Metal 金属板	Factory-coated white finish 工厂加白色装饰涂层	0.90	0.87	113

1.2 Roofs, Insulation Entirely above Deck

The insulation entirely above deck should be continuous insulation (c.i.) rigid boards. Continuous insulation is important because no framing members are present that would introduce thermal bridges or short circuits to bypass the insulation. When two layers of c.i. are used in this construction, the board edges should be staggered to reduce the potential for convection losses or thermal bridging. If an inverted or protected membrane roof system is used, at least one layer of insulation is placed above the membrane and a maximum of one layer is placed beneath the membrane.

1.3 Walls – Mass

Mass walls are defined as those with a heat capacity exceeding 7 Btu/ft^2·°F. Insulation may be placed either on the inside or the outside of the masonry wall. When insulation is placed on the exterior, rigid c.i. is recommended. When insulation is placed on the interior, a furring or framing system may be used, provided the total wall assembly has a U-factor that is less than or equal to the appropriate climate zone construction. See Figure Example Mass Wall Assembly.

1.2 楼顶上方的整体绝缘屋顶

楼顶上方的整体绝缘应该为连续的刚性绝缘物。连续绝缘至关重要，因为这样就能避免框架构件所导致的热桥效应或绝缘短路。当建筑采用两层连续绝缘层时，板层边缘应当错列开，以减少潜在的对流损失或热桥效应。如果采用反转或保护膜屋顶系统，至少应当有一层绝缘层位于保护膜之上，而保护膜下方至多只能有一层绝缘层。

1.3 大体块墙壁

大体块墙壁的热容量要超过7Btu/ft^2·°F（英热/平方英尺·华氏度）。砌筑墙的绝缘设在内外两侧均可。采用外侧绝缘时，建议使用连续的刚性绝缘物。采用内侧绝缘时，可能需要采用衬里或框架系统，前提是整个墙体的传热系数小于或等于相应气候带的适宜施工标准。见图，墙壁组合示例。

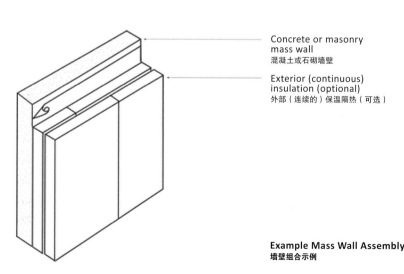

Concrete or masonry mass wall
混凝土或石砌墙壁

Exterior (continuous) insulation (optional)
外部（连续的）保温隔热（可选）

Example Mass Wall Assembly
墙壁组合示例

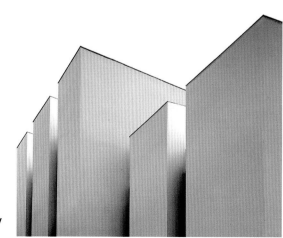

The greatest advantages of mass can be obtained when insulation is placed on its exterior. In this case, the mass absorbs heat from the interior spaces that is later released in the evenings when the buildings are not occupied. The thermal mass of a building (typically contained in the building envelope) absorbs heat during the day and reduces the magnitude of indoor air temperature swings, reduces peak cooling loads, and transfers some of the absorbed heat into the night hours. The cooling load can then be covered by passive cooling techniques (natural ventilation) when the outdoor conditions are more favourable. An unoccupied building can also be pre-cooled during the night by natural or mechanical ventilation to reduce the cooling energy use. This same effect reduces heating load as well.

Thermal mass also has a positive effect on thermal comfort. High-mass buildings attenuate interior air and wall temperature variations and sustain a stable overall thermal environment. This increases thermal comfort, particularly during mild seasons (spring and fall), during large air temperature changes (high solar gain), and in areas with large day-night temperature swings.

A designer should keep in mind that the occupant will be the final determinant on the extent of the usability of any building system, including thermal mass. Changing the use of internal spaces and surfaces can drastically reduce the effectiveness of thermal storage. The final use of the space must be considered when making the heating and cooling load calculations and incorporating possible energy savings from thermal mass effects.

1.4 Walls – Steel Framed

Cold-formed steel framing members are thermal bridges to the cavity insulation. Adding exterior foam sheathing as c.i. is the preferred method to upgrade the wall thermal performance because it will increase the overall wall thermal performance and tends to minimise the impact of the thermal bridging.

Alternative combinations of cavity insulation and sheathing in thicker steel-framed walls can be used, provided that the proposed total wall

对墙体外部进行绝缘可使利益最大化。墙体可以从室内空间吸热，并在夜晚建筑无人时释放出去。建筑的蓄热体（通常含在建筑外壳里）在白天吸收热量、减少室内空气温度的变化幅度、较少峰值制冷负荷，并且在夜晚将吸收的热量进行转移。这样一来，当室外条件允许时，被动式制冷技术（自然通风）就能覆盖制冷负荷。未投入使用的建筑同样可以在夜间通过自然或机械通风进行预制冷，以减少制冷的能耗。这同样能减少采暖负荷。

蓄热体还对热舒适度有积极的效果。大体量建筑能够减弱室内空气和墙面温度的变动，保持一个稳定的整体热环境。这可以增加热舒适度，特别是在气候温和的季节（春季和秋季）、气温变化剧烈（高太阳辐射）以及昼夜温差大的地区。

设计师应当考虑到建筑的居住、使用者是建筑使用价值（其中包括蓄热）的决定因素。改变室内空间和建筑表面的用途可能会彻底减弱蓄热体的效果。设计师在进行采暖和制冷负荷计算以及为建筑添加蓄热节能效果时，必须考虑到空间的最终用途。

1.4 墙壁的钢架结构

冷弯型钢框格是连接中空绝缘的热桥。在室外添加泡沫包层作为连续绝缘物可以优化墙体的热性能，因为它将增加墙体的整体热性能、将热桥效应最小化。

此外，也可以在较厚的钢架结构墙壁上选择采用中空绝缘和包板的混合模式，前提是整个墙体装配部件的

Recommendations on Envelope 建筑外壳设计建议

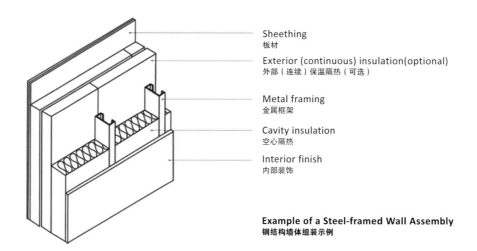

Example of a Steel-framed Wall Assembly
钢结构墙体组装示例

- Sheething 板材
- Exterior (continuous) insulation(optional) 外部（连续）保温隔热（可选）
- Metal framing 金属框架
- Cavity insulation 空心隔热
- Interior finish 内部装饰

assembly has a U-factor that is less than or equal to the U-factor for the appropriate climate zone construction. Batt insulation installed in coldformed steel-framed wall assemblies is to be ordered as "full width batts" and installation is normally by friction fit. Batt insulation should fill the entire cavity and not be cut short. See Figure Example (Above) of a Steel-framed Wall Assembly

传热系数小于或等于相应气候带建筑建议的传热系数。冷弯型钢框架墙壁的板状绝缘材料必须是"全宽"，通常以紧密摩擦进行安装。板状绝缘材料应当填满整个中空结构，不能被切短。见图，钢结构墙体组装示例。

1.5 Below-grade Walls

Insulation, when recommended, may be placed either on the inside or the outside of the below-grade wall. If placed on the exterior of the wall, rigid c.i. is recommended. If placed on the interior of the wall, a furring or framing system is recommended, provided the total wall assembly has a C-factor that is less than or equal to the appropriate climate zone construction.

1.5 地下墙壁

建议将绝缘层设在地下墙壁的内侧或外侧。如果设在墙壁外侧，建议使用连续的刚性绝缘物。如果设在墙壁内侧，建议使用衬里或框架系统，前提是整个墙体装配部件的传热系数小于或等于相应气候带建筑建议的传热系数。

1.6 Floors – Mass

Insulation should be continuous and either integral to or above the slab. This can be achieved by placing high-density extruded polystyrene above the slab with either plywood or a thin layer of concrete on top. Placing insulation below the deck is not recommended due to losses through any concrete support columns or through the slab perimeter.

Exception: Buildings or zones within buildings that have durable floors for heavy machinery or equipment could place insulation below the deck.

1.6 大面积地面

绝缘应当是连续的，可与楼板结合在一起，也可设在楼板上方。可以在楼板上方放置高密度挤塑聚苯乙烯，顶部铺上胶合板或薄层混凝土。不建议将绝缘设在楼板下方，因为混凝土支柱或楼板四周会散热。

特殊情况：建筑或建筑中的部分区域如果其地面十分坚固，适用于摆放沉重的设备或机械，可以采用楼板下方绝缘。

1.7 Floors – Metal Joist or Wood Joist/Wood Frame

Insulation should be installed parallel to the framing members and in intimate contact with the flooring system supported by the framing member in order to avoid the potential thermal short-circuiting associated with open or exposed air spaces. Non-rigid insulation should be supported from below, no less frequently than 24 in.

1.7 地面的金属托梁或木托梁/木结构

绝缘层应当与框架构件平行，通过框架构件的支撑于地面系统紧密相连，以避免与开放或外露的空气空间相关联的潜在热短路问题。不应在下方进行刚性绝缘，绝缘层不应超过24英寸（约0.61米）。

1.8 Doors – Opaque, Swinging

A U-factor of 0.37 corresponds to an insulated double-panel metal door. A U-factor of 0.61 corresponds to a double-panel metal door. If at all possible, single swinging doors should be used. Double swinging doors are difficult to seal at the centre of the doors unless there is a centre post.

1.8 不传热转门

双层绝缘金属门的传热系数为0.37。双层金属门的传热系数为0.61。在条件允许的情况下，应当尽量使用单开转门。如果中间没有立杆，双开门的中缝不便于密封。除了在对宽度有要求的区域，应该尽量少使用

Double swinging doors without a centre post should be minimised and limited to areas where width is important. Vestibules or revolving doors can be added to further improve the energy efficiency.

1.9 Doors – Opaque, Roll-Up, or Sliding

Roll-up or sliding doors are recommended to have R-4.75 rigid insulation or meet the recommended U-factor. When meeting the recommended U-factor, the thermal bridging at the door and section edges is to be included in the analysis. Roll-up doors that have solar exposure should be painted with a reflective paint (or should be high emissivity) and should be shaded. Metal doors are a problem in that they typically have poor emissivity and collect heat, which is transmitted through even the best insulated door and causes cooling loads and thermal comfort issues.

2. CAUTIONS

The design of building envelopes for durability, indoor environmental quality, and energy conservation should not create conditions of accelerated deterioration or reduced thermal performance or problems associated with moisture, air infiltration, or termites.

The following cautions should be incorporated into the design and construction of the building.

2.1 Slab Edge Insulation

Use of slab edge insulation improves thermal performance, but problems can occur in regions that have termites.

2.2 Air Infiltration Control

The building envelope should be designed and constructed with a continuous air barrier system to control air leakage into or out of the conditioned space and should extend over all surfaces of the building envelope (at the lowest floor, exterior walls, and ceiling or roof).

An air barrier system should also be provided for interior separations between conditioned space and space designed to maintain temperature or humidity levels that differ from those in the conditioned space by

没有中杆的双开门。通道或十字形旋转门的添加可以进一步提升能源效率。

1.9 不传热卷帘门或拉门

建议在卷帘门或拉门上采用R-4.75刚性绝缘或复合推荐传热系数的绝缘。在复合推荐传热系数的前提下，门和门框边缘的热桥效应可纳入分析之中。受阳光照射的卷帘门应当涂抹反射漆（或者材料本身应有高反射率），并且应当采用遮阳设施。金属门的问题比较严重，它们的反射率通常较低，且容易吸热。即使门的绝缘设计良好，也会导热。因此，金属门容易导致制冷负荷过高和热舒适度的问题。

2. 注意事项

为提升耐久度、室内环境质量及节能所进行的建筑外壳设计不应加速建筑的老化、降低其热性能或造成潮湿、空气渗透、蚁害等问题。

以下是建筑设计和施工过程中的注意事项。

2.1 楼板边缘绝缘

楼板边缘绝缘可以提高建筑的热性能，但是可能产生部分地区的白蚁蚁害问题。

2.2 空气渗透的控制

建筑外壳的设计和施工应当配有连续的气障系统来控制空气进入或泄露出空调调节的空间。气障系统应当延伸到建筑外壳表层的所有区域（最下层楼面、外墙、天花板或屋顶）。

在空调空间和温度或适度水平需要与空调空间保持超过50%差异度的空间之间的隔断，也应当使用气障系统。如果条件允许，应当在建筑内部使用鼓风机来寻

Recommendations on Envelope 建筑外壳设计建议

more than 50% of the difference between the conditioned space and design ambient conditions. If possible, a blower door should be used to depressurise the building to find leaks in the infiltration barrier. The air barrier system should have the following characteristics.
• It should be continuous, with all joints made airtight.
• Air barrier materials used in frame walls should have an air permeability not to exceed 0.004 cfm/ft^2 under a pressure differential of 0.3 in. H$_2$O (1.57 lb/ft^2) when tested in accordance with ASTM E 2178.
• The system should be able to withstand positive and negative combined design wind, fan, and stack pressures on the envelope without damage or displacement and should transfer the load to the structure. It should not displace adjacent materials under full load.
• It should be durable or maintainable.
• The air barrier material of an envelope assembly should be joined in an airtight and flexible manner to the air barrier material of adjacent assemblies, allowing for the relative movement of these assemblies and components due to thermal and moisture variations, creep, and structural deflection.
• Connections should be made between:
a. Foundation and walls
b. Walls and windows or doors
c. Different wall systems
d. Wall and roof
e. Wall and roof over unconditioned space
f. Walls, floors, and roof across construction, control, and expansion joints
g. Walls, floors, and roof to utility, pipe, and duct penetrations
• All penetrations of the air barrier system and paths of air infiltration/exfiltration should be made airtight.

找隔气层的漏点。
气障系统应当具备以下特征。
* 它应当是连续的，每个连接处都应密封好
* 依据美国试验材料协会的测试，在压差为0.3 in.（英尺）H$_2$O的情况下，框架墙的气障材料透气度不能超过0.004cfm/ft^2（立方英尺分/英尺）
* 系统应当能够抵挡建筑外壳上积极和消极的设计风速、电扇和烟囱压力，不会造成损坏或移位。系统还应传递结构的负荷，不应在全负荷状况下替代相邻的材料
* 建筑外壳组件上的气障材料应当给以密封和弹性的方式与相邻组件的气障连接好，保障这些组件的由于热度和湿度变化、蠕变、结构偏差所引起的相对运动
* 在以下结构之间应进行连接：
a.地基和墙壁；
b.墙壁和门窗；
c.不同墙壁系统之间；
d.墙壁和屋顶；
e.墙壁和无空调空间上方的屋顶；
f.墙壁、地面和跨结构、控制、伸缩接缝的屋顶；
g.墙壁、地面和与公共设施、管道、管道渗透相连的屋顶。
*所有气障系统的渗透点和空气内渗/外渗路径都应得到密封

Excerpted from Advanced Energy Design Guide for Small Hospitals and Healthcare Facilities © 2009 American Society of Heating, Refrigerating and Air-Conditioning Engineers, Inc. (www.ashrae.org)

摘录自小型医院与卫生保健设施先进节能设计指南©美国采暖、制冷与空调工程师学会(www.ashrae.org)

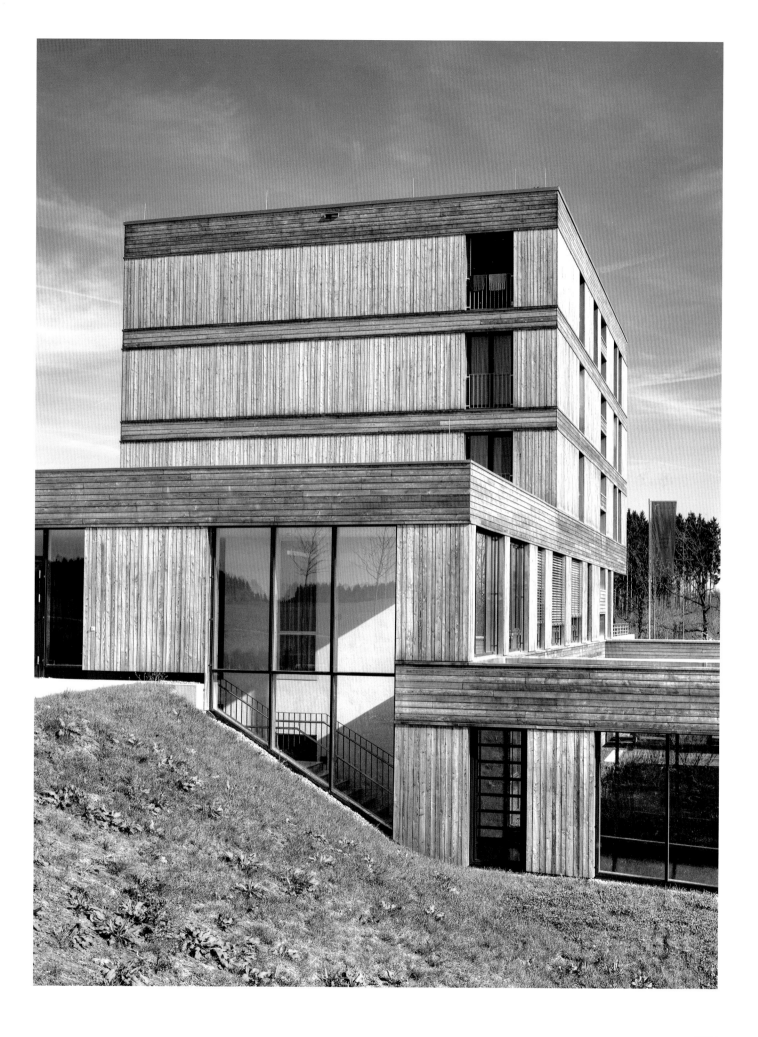

Recommendations on Windows
窗户设计建议

1. WINDOW DESIGN GUIDELINES FOR THERMAL CONDITIONS

Uncontrolled solar heat gain is a major cause of energy use for cooling in warmer climates and thermal discomfort for occupants. Appropriate configuration of windows according to the orientation of the wall on which they are placed can significantly reduce these problems.

1.1 Unwanted Solar Heat Gain is Most Effectively Controlled on the Outside of the Building

Significantly greater energy savings are realised when sun penetration is blocked before it enters the windows. Horizontal overhangs at the top of the windows are most effective for southfacing façades and must continue beyond the width of the windows to adequately shade them (see Figure below). Vertical fins oriented slightly north are most effective for east- and west-facing façades. Consider louvered or perforated sun control devices, especially in primarily overcast and colder climates, to prevent a totally dark appearance in those environments.

1. 窗户热状况设计指南

未受控制的太阳辐射热获得是炎热天气时造成制冷能耗和建筑内人员热舒适度降低的主要原因。根据墙面朝向进行合适的窗户配置能够有效解决此类问题。

1.1 在建筑外部可以有效地控制不必要的太阳辐射热增量

如果将太阳渗透阻挡在建筑外立面之外，就能够节约大量的能源。对朝南的外墙来说，窗户顶部的水平窗檐最为有效，窗檐宽度必须超过窗户的宽度，以便能够充分对其进行遮阳（详见下图）。微微向北倾斜的垂直扇片对东西两面的外墙最为有效。阴天和寒冷气候可以考虑采用百叶窗或穿孔太阳控制设备，以防止室内环境过于昏暗。

$$\text{Projection Factor} = \frac{\text{Horizontal Projection}}{\text{Height Above Sill}}$$

$$\text{投射因素} = \frac{\text{水平投射}}{\text{窗台以上的高度}}$$

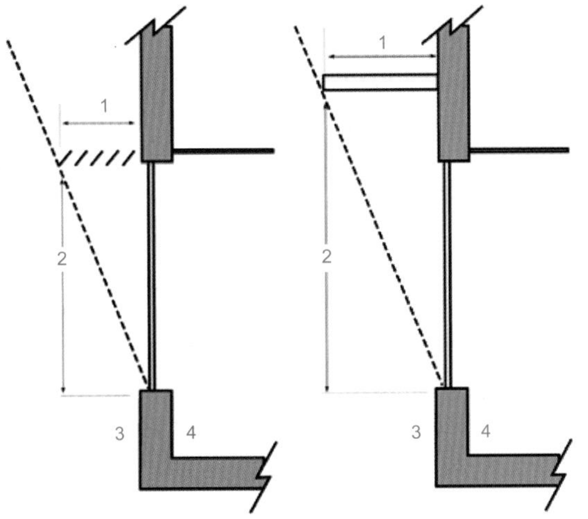

Windows with Overhang:
1. Projection
2. Height
3. Outside
4. Inside

有悬挑的窗体：
1. 投射
2. 高度
3. 外部
4. 内部

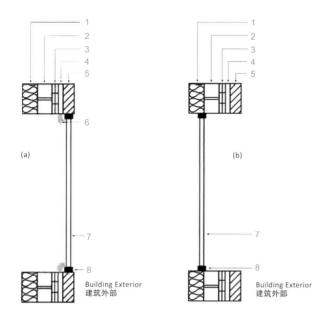

Thermal Break (a) at Window Frame and (b) in Window Frame Aligned with Wall Insulation:
1. Insulated Stud Space
2. Block
3. Foam Insulation
4. Air Gap
5. Brick
6. Frost or Condensation
7. Glazing
8. Frame thermal break

窗框（a）断热以及（b）窗框与墙隔热层之间的断热：
1. 隔热处理后的壁骨空间
2. 块料
3. 泡沫保温
4. 气缝
5. 砖
6. 结霜或冷凝
7. 施釉
8. 框架断热

1.2 Operable versus Fixed Windows

Operable windows play a significant role in healthcare design in embracing the core idea of indoor to outdoor connection and reaching out to the outdoor environment and nature.

However, although operable windows offer the advantage of personal comfort control and beneficial connections to the environment, individual operation of the windows not in coordination with the HVAC system settings and requirements can have extreme impacts on the energy use of a building's system. Advanced energy buildings with operable windows should strive for a high level of integration between envelope and HVAC system design. First, the envelope should be designed to take advantage of natural ventilation with well-placed operable openings. Second, the mechanical system should use interlocks on operable windows to ensure that the HVAC system responds by shutting down in the affected zone if the window is opened. The window interlock zones need to be designed to correspond as closely as possible to the HVAC zone affected by the open window.

Operable Clerestory Windows for Free Nighttime Cooling (Night Flush). In some cases operable windows may be used to remove thermal loads that have accumulated over the course of daytime and are stored in the building by cross-ventilating the building after business hours. Occupancy types best suited are administrative work areas, public spaces, and in some cases exam rooms. To allow for this to happen, the following conditions are required:
- Footprint: narrow floorplate and open-plan layout
- Operable windows
- Solid slabs/exposed ceiling slabs in concrete structures

1.3 Continuous Insulation to Avoid Thermal Breaks

Windows that are installed out of the plane of the wall insulation are one common source of envelope thermal breaks or breaches. Figure (Next page) shows an example of this construction. Installing the fenestration outside of the plane of the wall insulation defeats the thermal break in the window frame. In cold climates this causes condensation and frosting. The normal solution is not to rebuild the wall but to blow hot air against

1.2 可控窗与固定窗

可控窗在医疗设计中起到了重要的作用，它将室内外连接起来，使人们可以接触户外空间和大自然。

然而，尽管可控窗能够让人更舒适地进行控制并有利于接触自然环境，如果个人对窗户的控制与空调系统的设定和要求不一致，将会对建筑系统的能源效率起到负面的影响。配有可控窗的节能建筑应当对建筑外壳和空调系统设计进行高度的整合。首先，建筑外壳应当通过合理设置可控窗来充分利用自然通风。其次，机械系统应当运用可控窗上的联动装置来保证当窗户开放时，该区域的空调系统能够自动关闭。窗户联动装置区的设置应当尽量靠近受开窗影响的空调区。

利用可控天窗进行夜间制冷。在一些案例中可控窗可能被用来移除白天所累积以及工作时间后建筑通过自然通风所存储的热负荷。最适合此类制冷的区域包括行政工作区、公共空间和一些检查室。实现夜间制冷需要以下条件：
* 面积：狭窄的楼板和开放式布局
* 可控窗
* 混凝土结构的实心楼板或天花板外露楼板

1.3 用连续绝缘来避免热隔断

安装在墙板外部的窗户是形成热隔断或热裂效应的常见方式。下页图显示了此结构。将窗口安装在墙面绝缘板之外将会使窗框的热隔断失效。这在寒冷天气会造成冷凝和起雾。通常的解决方案是让热气冲向窗户，以提高室内玻璃和窗框表面的温度，通过增大内外玻璃的温差使室内膜层散热系数从0.68降到0.25。

the window to increase the interior surface temperature of the frame and glazing, which increases the temperature difference across the glazing and reduces the interior film coefficient from 0.68 to 0.25.

Fenestration should be installed to align the frame thermal break with the wall thermal barrier. This will minimise thermal bridging of the framing due to fenestration projecting beyond the insulating layers in the wall.

2. WINDOW DESIGN GUIDELINES FOR DAYLIGHTING

2.1 Visual Transmittance

Utilizing daylight in place of electrical lighting significantly reduces the internal loads and saves cost on lighting and cooling power. In the US, it is estimated that 10% of the total energy generated in 24 hours is consumed by electrical lighting during daytime. The higher the visible transmittance (VT), the more energy can be saved.

The amount of light transmitted in the visible range affects the view through the window, glare, and daylight harvesting. For the effective utilisation of daylight, high VT glazing types (0.60 to 0.70) should be used in all occupied spaces.

窗户的安装应当使热隔断装置和隔热层平行。这会最大限度地减少由于窗户凸出于墙体绝缘层之外所形成的框架热桥效应。

2. 窗户自然采光设计指南

2.1 可见光透射率

利用日光替代电力照明能够大幅减少建筑内部负荷，节约照明和制冷所消耗的电能。在美国，24小时内10%的总能源消耗来自于日间电力照明。可见光透射率越高，节约的能源就越多。

可见光透射的数量能影响透过窗户看到的景象、刺眼的光线和日光的收集。为了有效利用日光，应当在楼内区域采用具有高可见光透射率（0.60~0.70）的玻璃。

High VT's are preferred in predominantly overcast climates. VTs below 0.50 appear noticeably tinted and dim to occupants and may degrade luminous quality. However, lower VTs may be required to prevent glare, especially on the east and west facades or for higher WWRs. Lower VTs may also be appropriate for other conditions of low sun angles or light-coloured ground cover (such as snow or sand), but adjustable blinds should be used to handle intermittent glare conditions that are variable.

High continuous windows are more effective than individual ("punched") or vertical slot windows for distributing light deeper into the space and provide greater visual comfort for the occupants. Try to expand the top of windows to the ceiling line for daylighting but locate the bottom of windows no higher than 30 in. above the floor (for view). Daylighting can be achieved with higher WWRs, which can lead to higher heating and cooling loads.

2.2 Separating Views and Daylight

In some cases, daylight harvesting and glare control are not always best served by the same glazing product. Patient rooms in particular require better control of visual comfort levels, which can make it necessary to separate daylight glazing from view glazing.

The most common strategy is to separate (split) the window horizontally to maximise daylight penetration. For daylight glazing, which is located above the view window, between 6 ft above the floor and the ceiling, high VT glazing should be used. The view windows located below 6 ft do not

高可见光透射率的优势在阴天尤为明显。当可见光透射率低于0.50时，楼内人员会明显感到视野昏暗模糊，同时也会降低照明质量。然而，低可见光透射率可以避免刺眼的光线，特别适用于窗墙面积比较大的东西外墙。低透射率还适合太阳角度较低或地面颜色较浅（如雪或沙）的情况，但是应当使用可调节百叶窗来处理间歇性的刺眼光线。

连续的高窗比独立或垂直的窄窗效率更高，它们能将光线引入空间深处，提供更高的视觉舒适度。建议将窗顶延伸到天花板线进行日光照明，而窗户底部则不应高于30英寸（约0.76米）。窗墙面积比越高，日光照明效果越好，但是会导致采暖和制冷负荷增加。

2.2 分隔观景和采光

在一些案例中，同一种玻璃制品并不能完美地实现收集日光和控制刺眼光线两种功能。由于病房特别需要对视觉舒适度进行良好控制，有必要将病房内的采光窗和观景窗分开。

最常见的策略是将窗户水平隔离开，以实现日光渗透的最大化。位于观景窗上方的采光窗（高于地面约6英尺，相当于1.83米）应当采用高可见光透射率玻璃。而低于6英尺（约1.83米）的观景窗则不需要采用高透

require such high VTs, so values between 0.50 and 0.60 are acceptable to achieve recommended SHGC values.

Windows both for view and for daylighting should primarily be located on the north and south façades. Windows on the east and west should be minimised as they are difficult to protect from overheating and from glare.

2.3 Colour-Neutral Glazing
The desirable colour qualities of daylighting are best transmitted by spectrally neutral glass types that alter the colour spectrum to the smallest possible extent. Avoid tinted glass, in particular bronze- and green-tinted glazing.

2.4 Reflectivity of Glass
To the greatest extent possible, avoid the use of reflective glass or low-e coating with a highly reflective component. These reduce transparency significantly, especially at acute viewing angles where they impact the quality of the view.

2.5 Light-to-Solar-Gain Ratio
High-performance and selective low-e glazing permits significantly higher visual transmittance than reflective coatings or tints do. The light-to-solar-gain (LSG) ratio is the criteria for stating the efficacy of the glass, indicating the ability to maximise daylight and views while minimising solar heat gain. In today's markets a variety of costeffective glass types are available with high LSG ratios. Ratios over 1.6 are considered good. Any ratio greater than 2.0 is very effective and will contribute to achieving the goal of 30% savings.

2.6 High Ceilings
More daylight savings will be realised if ceiling heights are raised along the building perimeter. Greater daylight savings can be achieved by increasing ceiling heights to 11 ft or higher and by specifying higher VTs (0.60 to 0.70) for the daylight window than for the view window. North-facing clerestories are more effective than skylights to bring daylight into the building interior.

射率玻璃，透射率在0.50和0.60之间就能满足太阳的热系数的要求。

拥有观景和采光双重功能的窗户主要位于南北两个立面。应尽量减少东西两侧的窗户数量，避免过热和眩光现象的产生。

2.3 中性色彩的玻璃装配
中性玻璃对可见光色谱的改变极小，所传递的光线具有日光照明的理想色彩质量。应当避免使用彩色玻璃，特别是古铜色和绿色调的玻璃。

2.4 玻璃的反射率
尽量不要使用反射玻璃或带有高反射率的低辐射涂层。这些都会降低透明度，特别是在特定的角度会影响观景的质量。

2.5 光热比
高性能的低辐射玻璃比反射玻璃和彩色玻璃的可见光透射率要高得多。光热比是玻璃效力的重要指标，标志着玻璃将日光和视野最大化及将太阳热增量最小化的能力。当前市场上有多种具有良好性价比的高光热比玻璃类型。光热比高于1.6的玻璃性能就可被评为优良。光热比高于2.0的玻璃非常高效，能够实现节能目标的30%。

2.6 高天花板
将天花板的高度沿着建筑边缘升高能够让日光照明实现更显著的节能效果。如果将天花板高提升到11英尺（约3.35米）或更高，并且对采光窗使用更高可见光透射率（0.60~0.70）的玻璃，就能实现更多的能源节约。朝北的侧天窗比正天窗更高效，能为建筑室内带来更多的日光。

2.7 Light Shelves

Consider using interior or exterior light shelves between the daylight window and the view window. These are effective for achieving greater uniformity of daylighting and for extending ambient levels of light onto the ceiling and deeper into the space. Some expertise and analysis will be required to design an effective light shelf.

2.7 遮阳板

考虑在采光窗和观景窗之间使用室内或室外遮阳板。这可以提高日光照明的统一度，使光线延伸到天花板和空间的深处。设计高效的遮阳板需要一些专业知识和分析。

Recommendations on Lighting

照明设计建议

Energy-efficient lighting in the healthcare setting is possible without sacrificing the visual needs and comfort of patients and caregivers. There are also potentially beneficial health effects of exposing patients and staff to daylight in many settings. When properly designed, a positive synergy of energy conservation and improved health is possible.

在不影响患者和医务人员的视觉需求和舒适度的前提下，医疗设施也能实现节能照明。当然，让患者和工作人员每天多照射阳光还有许多潜在的健康益处。只要设计合理，节能和健康完全可以结合在一起。

1. ELECTRIC LIGHTING

1.1 Light-Coloured Interior Finishes

For electrical lighting to be most efficient, spaces must have light-colored finishes. Ceiling reflectance should be at least 85% for direct lighting schemes and preferably at least 90% for indirect and daylighting schemes. This generally means using high-performance white acoustical tile or high-reflectance white ceiling paint on hard surfaces. For daylighting schemes, the average reflectance of the walls should be at least 50%, and for the portions of the wall adjacent to the daylighting aperture and above 7 ft high it should be 70%. This generally means using light tints for the wall surface, as the lower reflectance typical of doors, trim, and other objects on the walls will reduce the average. Floor surface reflectance should be at least 20%, for which there are many suitable surfaces.

The shape and finish of the ceiling should also be considered. A flat or gradually sloped ceiling is the most efficient; steep sloping ceilings and exposed structures, even if painted white, may have significantly lower reflectance. Lighting systems with indirect components are recommended in some applications, but if the ceiling cavity includes exposed structures or exposed ductwork, a higher percentage of direct light may be required.

1.2 Linear Fluorescent Lamps and Ballasts

T-8 lamps with electronic ballasts are a very commonly specified commercial fluorescent lighting system in the United States. Fluorescent lamps with low mercury content are available from major lamp manufacturers and are the standard for sustainable design projects.

To evaluate the efficacy of the lighting system, consider the mean, or

1. 电气照明

1.1 浅色室内装饰

为了使电气照明更高效，室内必须采用浅色装饰。直接照明区域的天花板的反射率至少应为85%，而间接照明和日光照明区域的天花板反射率最好在90%以上。这意味着建议设计需要使用高性能白色隔音砖或在坚硬的表面上涂上高性能的白色天花板涂料。在日光照明区域，墙面的平均反射率至少应为50%，而与日光照明窗口相邻及高于7英尺（约2.13米）的墙壁的反射率至少应为70%。因此，建议在墙面上使用浅色涂料，因为门、镶边及墙面上的其他部件都会降低平均反射率。地面反射率至少应为20%，有多种选择。

应当考虑到天花板的造型和装饰。平面或缓坡面的天花板最为高效；陡坡面天花板和显式结构即使漆成了白色，其反射率也会很低。许多区域都可以使用间接照明系统，但是如果吊顶空间包含显式结构或露出了管道，则最好使用高百分比的直接照明。

1.2 荧光灯管和镇流器

配有电子镇流器的T-8荧光灯是美国商业设施照明系统的常见灯具。许多大型灯具制造商都提供低含汞量的荧光灯，而低含汞量的荧光灯正好符合可持续设计项目标准。

为了评估照明系统的效率，可以考虑灯具制造商在产品

"design," lamp lumens in the lamp manufacturer's specification data. This value is lower than the initial lumens and reflects the depreciated lumen output occurring at 40% of the lamp's rated life, which better characterises actual performance.

Generally, the lumen output will vary according to the colour temperature, colour rendering index (CRI), and between standard (RE700), premium (RE800), and high performance (RE800/HL) lamps. The light source efficacy and LPD requirements can be achieved as long as the higher performance versions of T-8 lamps and ballasts are used.

The CRI is a scale measurement identifying a lamp's ability, generally, to adequately reveal colour characteristics. The scale maximises at 100, which indicates the best colour-rendering capability. Lamps specified for ambient lighting should have a CRI of 80 or greater to allow for accurate perception of colour characteristics.

High-performance T-8 lamps are defined, for the purpose of this document, as having a lamp efficacy of 90+ lumens per watt, based on mean lumens divided by the catalogued lamp input watts. High-performance T-8s also have a CRI of 82 or higher and no less than 95% lumen maintenance.

Next, select the ballast. The ballast has significant impact on the energy efficiency of the lighting system. Similar to lamp efficacy, lighting system efficacy is a measure of the energy efficiency of the combined lamp and ballast system.

To determine the lighting system efficacy, which is expressed in mean lumens per watt (MLPW), multiply the lamp mean lumens by the number of lamps and the ballast factor (BF), and then divide by the ballast input power (watts). For example, using two standard T-8 lamps and a generic instant start ballast, the system efficacy is

说明书中标出的平均/设计流明数。这一数值低于初始流明数，反映了当灯管寿命超过40％以后的折旧流明输出，更能体现其实际性能。

总体来讲，流明输出会根据色温、色彩还原度而有所不同。标准灯管（RE700）、高级灯管（RE800）和高性能灯管（RE800/HL）的流明输出也不相同。只要采用T-8以上性能等级的灯管和镇流器，就能实现所要求的光源效力和照明功率密度。

色彩还原度是衡量灯具性能的标杆，显示其是否能够充分展示色彩特性。色彩还原度的最高值是100，表示最好的色彩还原能力。环境照明所使用灯具的色彩还原度应当为80及以上，以确保准确的色彩感觉。"标准"T-8灯的色彩还原度比推荐的规格要低。

以分类灯具的输入功率来计算平均流明，高性能T-8灯的效率可达每瓦90+流明。高性能T-8灯的色彩还原度可达到82及以上，并且能保持95％以上的流明数。

其次，选择镇流器。镇流器对照明系统的节能有着重要的影响。与发光效率相似，照明系统的效率是衡量灯管和镇流器系统能效的标准。

照明系统的效率（以流明每瓦（MLPW）来表现）的计算方法是用灯管平均流明数乘以灯管数量和流明系数（BF），然后除以镇流器的输入功率（W）。例如，使用两个标准T-8灯和一个通用即时启动镇流器，系统效率的计算公式如下：

$$\frac{2 \text{ lamps} \times 2660 \text{ mean lumens} \times 0.84 \text{ ballast factor}}{59 \text{ watts}} = 77 \text{MLPW}$$

$$\frac{2\text{灯泡} \times 2660\text{平均流明} \times 0.87\text{流明系数}}{59\text{瓦特}} = 77\text{照明系统效率（流明/瓦）}$$

Standard "Generic" Instant Start Electronic Ballasts. The most standard instant start ballast is the common and least expensive ballast; the typical input power for a twolamp normal light level (0.87 BF) is about 59 W. If you do not specify ballast performance, this is likely what the manufacturer will use in the luminaire.

Low Light Output Version of Standard Ballasts. Similar to the standard ballast, this version operates at 0.78 BF and has input power of about 54 W for a two-lamp ballast. The resulting light level is about 10% less than the standard ballast, with a corresponding reduction in input power.

High Light Output Version of Standard Ballasts. Similar to the standard ballast, this version operates at 1.15–1.20 BF and has input power of 74 to 78 W for a two-lamp ballast. The resulting light level is about 32% higher than the standard ballast, with a corresponding increase in ballast input power.

Program Start Ballasts. Available in low power and normal power models, program start ballasts use an additional watt per lamp to perform programmed starting, which makes lamps last longer when frequently switched.

Dimming Ballasts. Electronic ballasts that provide a continuous range of dimming are available in varying ranges from 100%–20% to 100%–1%. Most dimming ballasts require 60 to 66 W for two lamps. Additional power, compared to fixed output ballasts, is used to heat the lamp cathodes to permit proper dimming operation, but some newer high-performance dimming ballasts do have full output MLPW over 90. (MLPW efficacy is less valuable for evaluating dimming ballasts since the lumen output, BF, and corresponding input power vary over the dimming range of the ballast). Another variation is "stepped dimming," which typically provides 2 or 3 levels, such as 50% and 100% or 35%, 65%, and 100%.

High-Performance Electronic Ballasts. A high-performance electronic ballast is defined, for the purpose of this document, as a two-lamp ballast using 55 W or less with a BF of 0.87 (normal light output) or greater. High-performance ballasts are also available for low light output, high light output, and dimming versions.

标准通用即时启动电子镇流器。标准即时启动镇流器最为常见，价格也最低廉。两个标准光级度灯管的标准输出功率约为59瓦。如果不特别说明镇流器性能，这就是制造商在发光体上的常用标准。

标准镇流器的低光输出版本。与标准镇流器相似，此版本的流明系数在0.78，双灯管镇流器的输入功率约为54瓦。由于输入功率的缩减，所生成的光级度比标准镇流器少10%。

标准镇流器的高光输出版本。与标准镇流器相似，此版本的流明系数在1.15~1.20，双灯管镇流器的输入功率约为74瓦到78瓦。由于输入功率的增加，所生成的光级度比标准镇流器多32%。

程序起始镇流器。程序起始镇流器适用于低功率和正常功率模式，采用附加的功率来实现程序启动，能够延长经常开关的灯管的寿命。

调光镇流器。电子镇流器的连续调光范围在100%~20%和100%~1%之间。大多数双灯调光镇流器为60瓦–66瓦。与定输出镇流器相比，它的附加功率主要用于加热灯管的阴极来实现合适的调光操作。一些较新的高性能调光镇流器的输出照明效率可以超过90（由于流明输出、流明系数和相应的输入功率随着镇流器调光范围而不尽相同，调光镇流器的照明系统效率也不便衡量）。另一种变化为"阶梯性调光"。这种调光方式通常分为2到3级，如50%和100%，35%、65%和100%。

高性能电子镇流器。双灯镇流器功率小于55瓦，且流明系数可达到0.87（标准光输出）或更高的，即可被称为高性能电子镇流器。高性能镇流器同样有低光输出、高光输出、调光等版本。

1.3 Compact Fluorescent

To achieve appropriate LPD, compact fluorescent lamps (CFLs) can be used for a variety of applications such as utility lighting in small spaces, downlighting, accent lighting, and wall-washing. Suitable lamps include twin tube, multiple twin tube, twist tube, and long twin tube lamps. Only pin-based CFLs are included in this group, since a screw-based lamp can be replaced with an incandescent lamp and is therefore not compliant with most energy codes. Suitable luminaires have integral hard-wired electronic ballasts.

Because the efficacy of CFLs is typically less than 60 MLPW, they should not be used for general lighting in most space types. To meet the 50 MLPW efficacy requirements, some CFL-and-ballast combinations must be avoided.

1.4 Metal Halide

To achieve appropriate LPD, metal halide lamps may be used for general lighting in large spaces outdoor lighting, and for accent lighting and wall-washing in low wattages. In the metal halide family there are two primary types: ceramic metal halide (CMH) lamps and quartz metal halide (QMH) lamps. Both types are high-intensity discharge lamps in which intense light energy is generated inside an arc tube made either of ceramic or quartz glass. The two types are comparably efficient. CMH lamps have very good colour in the warm (3000 K) and neutral (4000 K) ranges; QMH lamps' colour rendering quality is mediocre except in high colour temperature lamps (5000 K and above). In general, only the improved CRI CMH lamps are recommended for interior applications.

Caution: Metal halide lamps require a warm-up and re-strike time of up to 15 minutes if turned off during operations. Therefore a supplemental emergency source is required that will provide light during the re-strike time when used in applications requiring emergency standby power. In addition, metal halide lamp performance is affected by the position of the lamp arc tube. When the lamp is operated in a position other than the rated position, the output will be reduced. This is known as the tilt factor. Lamps applied in a manner where tilt factor reduces output will have reduced efficacy that may fall outside the recommendations.

1.3 紧凑型荧光灯

为了实现照明功率密度标准，可以在多处应用紧凑型荧光灯，例如小空间的通用工具灯、下照灯、重点照明、墙面泛光照明等。合适的灯具有双联管、多重双联管、加捻管和长双联管。只有插入式紧凑型荧光灯符合这个分组，螺旋式灯可由白炽灯替代，因此没有一致的能源标准。合适的灯具配有内置硬连线电子镇流器。

由于紧凑型荧光灯的效率通常低于60流明/瓦，它们不适合用于普通照明。为了满足50流明/瓦的效率要求，尽量避免使用一些类型的紧凑型荧光能和镇流器。

1.4 金属卤素灯

为了实现照明功率密度标准，可以在大型户外空间、低功率重点照明和墙面泛光照明等处使用金属卤素。金属卤素主要有两种类型：陶瓷金属卤素灯和石英金属卤素灯。两种类型都是高强度气体放电灯，照明原理是在陶瓷或石英玻璃电弧管内释放光能。这两种类型都比较高效。陶瓷金属卤素灯在3,000K和4,000K之间的色彩较好；石英金属卤素灯的色彩还原度较为普通，只有在高色温（5,000K及以上）灯具中表现才更突出。总体来讲，室内应用只推荐使用改良的陶瓷金属卤素灯。

注意：如果在操作过程中关闭，金属卤素灯的预热和再启动时间可达15分钟。因此，当需要使用紧急备用电源时，必须保证有附加的紧急光源在卤素灯再启动期间提供光线。此外，金属卤素灯的性能会受到电弧管位置的影响。当卤素灯在非设计位置使用时，其输出率将会降低。这就是倾斜系数。当倾斜系数减少灯具的输出率时，灯具的效率可能无法达到推荐标准。

1.5 Exit Signs

Use light-emitting diode (LED) exit signs or other sources that use less than 5 W per face. The selected exit sign and source should provide the proper luminance to meet all building and fire code requirements.

1.6 General Lighting Control Strategies

To maximise the energy performance of a facility, lighting control strategies should be adopted to optimise how lights are turned on, when and how lights are turned off, and the output level of the lights whenever they are operating. The optimum electric light level in a given space is dependent on the task or activity in the space, the user's personal preference or desired aesthetic, and the amount of daylight in the space. Typically, the highest potential for lighting energy savings in a hospital is to adjust the lighting to optimum levels whenever the lights are turned on.

One example of this approach is the common need for relatively high illuminance levels in a space for housekeeping and maintenance where lower light levels would be appropriate for typical operation or patient care. In this application, having the higher level of illumination separately controlled by a manual ON-time interval OFF switch, instead of setting the room lighting at the higher level all the time, will save energy. Another prime light control strategy for most public spaces, corridors, waiting rooms, etc. is daylighting control, where electric light levels are automatically adjusted to supplement the available daylight in a space throughout the day.

1.7 Occupancy-Based Control

Use occupancy sensors in all exam and treatment rooms as well as in staff support spaces for nutritional care, medication areas, clean and soiled utility rooms, offices, mechanical rooms, restrooms, and storage rooms. The greatest energy savings and occupancy satisfaction are achieved with manual ON/automatic OFF occupancy sensor scheme (also referred to as vacancy sensing). This avoids unnecessary operation when electric lights are not needed, reduces the frequency of switching, and maximizes lamp life compared to automatic ON schemes. In every application the occupant should not be able to override the automatic OFF setting, even

1.5 出口标志

建议使用发光二极管（LED）出口标志或其他每面低于5瓦的光源标志。选定的出口标志和光源应当提供符合建筑和消防适用标准的光亮度。

1.6 通用照明控制策略

为了实现设施的能源性能最大化，需要采用照明控制策略来优化灯具的开启方式、灯具的关闭时间和关闭方式以及灯具使用时的输出水平。某个空间的最佳电灯照明水平取决于该空间的任务或所进行的活动、使用者的个人偏好或审美观、该空间的日光照射量等。通常来说，医院里最能实现照明能源节约的方式是调整灯光的最佳光级度。

以上策略的具体实践体现在：整理内务和进行维护的空间应采用较高的照明度，而普通手术和患者护理空间则应采用相对较低的照明度。这样一来，利用手动调光开关，而不是一直将空间的照明度保持在较高水平，就能实现节能。另一个常用于公共空间、走廊、候诊室等区域的照明控制策略是进行日光照明控制。电灯的照明度可以随着白天空间内的日光量进行自动调剂，补足所需的亮度。

1.7 感应控制

可以在所有检查室、治疗室、员工服务区、营养护理区、药物治疗区、杂物房、办公室、机房、洗手间和储藏室使用传感器。最节能而舒适的方式是采用手动开/自动关的感应控制系统（又称空缺传感）。这避免了电灯的不必要使用，减少了开关的频率，还延长了灯具的寿命（与自动开系统相比）。在任何情况下，使用者都不应改变自动关闭设置，即使在有手动开的情况下也不行。当某个空间只是偶尔需要采用高等级光照时，还可以考虑使用手动关或独立的调光设备。如无例外，传感

if it is set for manual ON. A manual OFF or separate switching capability is also useful for multi-level lighting schemes where a higher light level is only needed periodically. Unless otherwise recommended, occupancy sensors should be set for medium to high sensitivity and a 15-minute time delay (the optimum time to achieve energy savings without excessive loss of lamp life). Review the manufacturer's data for proper placement and coverage.

In high-performance integrated lighting control systems, motion sensors can also be used in spaces that have little if any traffic during late-night hours. If light levels are automatically set back late at night, motion sensors can be used to raise light levels in public corridors, waiting rooms, and other spaces whenever someone approaches and occupies the space.

The two primary types of occupancy sensor technologies are passive infrared (PIR) and ultrasonic. PIR sensors can see only in a line-of-sight and should not be used in rooms where the user cannot see the sensor (e.g., storage areas with multiple aisles, restrooms with stalls). Ultrasonic sensors can be disrupted by high airflow and should not be used near air duct outlets. Dual-technology sensors combine the sensor technologies; these should be considered for spaces larger than 150 ft^2 or those with objects or partitions that could affect the performance of PIR sensors. Sensors can also incorporate auxiliary relays that can interface between lighting and other building systems.

Caution: Occupancy sensors should not be used with high-intensity discharge (HID) lamps because of warm-up and re-strike times. Programmed start ballasts are recommended for fluorescent lamps and CFLs if frequent on-off cycles are expected. (Some standard ballasts and all dimming ballasts are programmed start for some lamp types.)

1.8 Daylight Harvesting Control

In atriums, lobbies, waiting rooms, corridors, open-office and administration areas, and other appropriate spaces where the design team has brought quality daylight into the space, automated daylight harvesting controls can be used to regulate the output of electric lights to optimise the quality

器应当被设置在中度到高度敏感度，延时时长为15分钟（这是既能实现节能又不会损害灯具寿命的最佳时长）。通过制造商提供的数据查阅灯具的合适设置和覆盖范围。

在高性能综合照明控制系统中，还可以在深夜有人经过的区域采用运动传感器。如果夜晚的光照度被自动设为较低，可以利用运动传感器来提高公共走廊、候诊室以及其他有人使用空间的亮度。

传感器控制技术有两种主要类型，分别为被动红外线和超声波。被动红外传感器只能感应视线的范围，因此不宜在使用者看不到传感器的房间内使用（例如：走道较多的仓储区域、有隔间的洗手间）。超声波传感器可能会被高速气流所影响，不宜安装在排风口附近。双重技术传感器集合了两种技术，适用于面积大于150平方英尺（约13.94平方米）的空间以及有物体或隔断影响红外传感器效果的空间。传感器还可以包含辅助继电器，将照明系统与其他建筑系统连接起来。

注意：传感器不能与高强度放电灯一起使用，因为后者有预热和再启动时间。如果存在频繁的开关，建议在荧光灯上使用程序启动镇流器（部分标准镇流器和全部调光镇流器对部分灯型都可采用程序启动）。

1.8 日光收集控制

在中庭、大堂、候诊室、走廊、开放式办公区、行政区域以及其他设计团队认为适合采用日光照明的空间都可以采用自动的日光收集控制来调节电灯的输出，优化视觉环境的质量，节约大量能源。在电灯照明陡

of the visual environment while saving significant amounts of energy. Step dimming systems can be applied where abrupt incremental changes in ambient electric light levels will not be a distraction to the occupants in the space. In spaces where adjustments in electric light levels should be transparent to the occupants, continuous dimming systems should be applied.

Daylight harvesting controls may be considered in patient room applications, especially for the lighting zones nearest the windows. Lighting power reductions during daylight hours have been as high as 87% in this application (Brown et al. 2005). However, patient control of their environment is an overriding priority, and automatic controls should not override the ability to manually control the lights. This makes the potential energy savings challenging to quantify.

1.9 Electrical Lighting Design
The 1.0 W/ft² LPD represents an average LPD for the entire building. Alternatively, individual spaces may have higher power densities if they are offset by lower power densities in other areas. In this space-byspace method, calculate an overall lighting power allowance for the entire project by adding the products of the individual space type areas.

2. DAYLIGHTING

2.1 General Principles
Daylighting is not limited to specific technologies. Daylighting is based on an integrated approach to design that takes influence at every scale and level of design and during each phase of the design process.

Daylighting strategies drive building shape and form, integrating them well into the design from structural, mechanical, electrical, and architectural standpoints.

Daylighting increases energy performance and impacts building size and costs by downsizing fans, ductwork, and cooling equipment because overall cooling loads are reduced, allowing for trade-offs between the

然变化不会影响使用者的情况下，可以采用阶梯式调光系统。当使用者需要无感觉的调光时，应当采用连续调光系统。

日光收集控制可以应用于病房，特别是靠近窗户的区域。此种应用在日照时间所减少的照明耗电可高达87%（布朗等，2005）。然而，患者对自己环境的控制应当被放在第一位，自动控制不应限制手动控制系统。这使节能效果的量化变得较为困难。

1.9 电气照明设计
1.0 W/ft²的照明功率密度代表了整座建筑的平均照明功率密度。相对的，个别空间的功率密度可能较高，而其他空间的功率密度较低，二者可以相互抵消。用这种逐个空间的计算方式，通过叠加独立空间区域的产品和它们所代表的照明功率密度。

2. 日光照明

2.1 总则
日光照明不仅限于特定的技术。日光照明以综合设计为基础，涉及设计的各个层面以及设计流程的各个阶段。

日光照明策略能影响建筑的造型，使其在结构、机械、电气和建筑方面与设计相结合。

日光照明能提高能源性能，通过减少风扇、管道以及制冷设备的数量（因为整体制冷负荷被降低，可以权衡日光照明的成本，调节空气处理和制冷系统的规模）来缩

efforts made for daylighting and the sizing of the air-handling and cooling systems.

Providing daylight is fundamental for a healing environment, as it makes a key contribution to energy-efficient and eco-effective healthcare design. While the most valuable asset of daylight is its free availability, the most difficult aspect is its controllability as daylight changes during course of the day. Daylighting is more of an art than a science, and it offers a broad range of technologies that provide glare-free balanced light, sufficient lighting levels, and good visual comfort.

Daylighting will only translate into savings when electrical lighting is dimmed or turned off and is replaced with natural daylight.

Effective daylighting uses natural light to offset electrical lighting loads. When designed correctly, daylighting lowers energy consumption and reduces operating and investment costs:
• Reduced electricity use for lighting and peak electrical demand
• Reduced cooling energy and peak cooling loads
• Reduced fan energy and fan loads
• Reduced maintenance costs associated with lamp replacement
• Reduced HVAC equipment and building size and cost

However, to achieve this reduced cooling, the following criteria must be met:
• High-performance glazing to meet lighting design criteria and block solar radiation
• Effective shading devices, sized to minimize solar radiation during peak cooling times
• Electric lights, through the use of photosensors, automatically dimmed or turned off

The case for daylighting reaches far beyond energy performance alone. Indoor environmental quality not only benefits the patients and their healing process, but also has a significant impact on the performance of the care-giving staff.

减建筑规模和建筑成本。

对治疗环境来讲，日光照射至关重要，因为它是医疗设计中实现节能环保的主要策略。日光最大的优势在于它的免费，但它的劣势在于不可控制。日光照明更像一门艺术，而不是一门科学。它运用大量的技术来提供不刺眼的平衡光、足够的照明度以及良好的视觉舒适度。

只有当电气照明被调暗或关闭，以自然光替代照明时，日光照明才能被换算成节约的能源。

高效的日光照明利用自然光来抵消电气照明负荷。合理的设计能够让日光照明降低能耗，从而减少运营和投资成本：
* 减少照明用电和峰值电力需求
* 减少制冷能耗和峰值制冷负荷
* 减少风扇能耗和风扇负荷
* 减少与灯具相关的维护费用
* 减少空调系统设备的使用，缩减建筑规模和成本

然而，为了实现缩减制冷，必须实现以下条件：
* 利用高性能玻璃装配以满足照明设计要求并阻挡太阳辐射
* 利用高效遮阳设备在峰值制冷时间最小化太阳辐射量
* 利用感光器使电灯自动调暗或关闭

日光照明不仅能提升能源绩效，良好的室内环境质量不仅对患者和他们的康复过程有益，还能积极地影响医护人员的工作表现。

Daylight is an essential component for improving patient recovery and for reducing the patient's time of stay. The most underestimated value of daylighting is its ability to increase staff productivity and reduce medical errors. These impacts are difficult to quantify, but the potential for improvement and economical savings is immense and needs to be taken into consideration as serious decision making criteria in the process of healthcare design. These benefits may far outweigh the energy savings and become the significant drivers for daylighting buildings altogether.

The daylighting strategies recommended in this guide have successfully been implemented in buildings before. Most daylighting strategies are generic and apply to healthcare facilities just as they do to other building types. The reason daylighting has been implemented less successfully in healthcare settings is due to programmatic constraints that make it more challenging to locate occupied spaces on the perimeter than it is for other building types. Also, occupancy-specific lighting conditions, as required for patient rooms, reduce opportunities for daylighting to specific space types only. The following tips and strategies are designed to address healthcare-specific opportunities and to overcome these obstacles.

2.2 Consider Daylighting Early in the Design Process

In small healthcare facilities, the building programme and medical planning are the main drivers that establish the shape and the footprint of the building. Planning criteria often result in creating compact, deep floor plates, while daylighting strategies attempt the opposite by articulating and narrowing the floor plate.

The configuration of the building footprint is established early in the design process, definitely freezing the building depth early and locking in all future potential for daylighting – the key factor for anticipating future design upgrades and improvements. A frequent issue with existing buildings is their depth of floor plate, which prevents easy upgrades with daylighting and natural ventilation.

This demonstrates two important aspects. One is the importance of integrating daylight design criteria before the footprint is locked in so that the building can unfold its full energy-saving potential. Another is

日光对促进患者康复、缩短其住院时间有着积极的作用。日光最易被忽视的价值就是它能够增加工作人员的工作效率，减少医疗失误。这些影响难以量化，但是它们的改良效果和经济价值是巨大的，应当在医疗设计之中占有一席之地。这些好处可能会远远超出日光的节能价值，成为设计日光照明建筑的主要驱动力。

本指南所推荐的日光照明策略已经在以往的建筑实践中得到了验证。大多数适用于其他建筑类型的日光照明策略也都适用于医疗设施。之所以医疗设施很少采用成功的日光照明是因为它的设计有许多限制，很难像其他建筑类型一样将日常使用的空间都设计在建筑外围。此外，病房针对患者所设计的照明条件缩小了日光照明的可行范围。以下建议和策略专为医疗设施的日光照明所设计，可以帮助克服以上难题。

2.2 在设计初期考虑日光照明

在小型医疗设施的设计中，建筑方案和医疗规划是决定建筑造型和轮廓的两大主要因素。根据设计标准通常会设计出紧凑而纵深较深的楼面，而日光策略则要求有狭长的楼面。

建筑轮廓的配置在设计初期就已经决定，会在早期就固定建筑的纵深，阻断日光照明的可能性，使设计在未来无法升级或改进。对现有建筑来讲，阻止建筑进行日光照明和自然通风改造的通常都是它们的楼面纵深。

这说明了两个重要的方面。一是在确定建筑轮廓之前决定日光设计的重要性，以便建筑能够实现其全部节能潜力。二是医疗规划和能效设计是不可分割的两部分，共

that medical planning and energy-efficient design are inseparable design criteria, as they both impact the shape and footprint and are integral drivers of the shaping of the "bones" of the building.

Daylight strategies impact the design at different levels of scale in each phase of design and can be characterized in four categories.

Pre-Design. During pre-design, the daylight strategies' focus is on massing studies and the shaping of the floor plate. The goal is to minimise depth and maximise access to windows and daylight by strategically placing light wells, shafts, and atria and orienting fenestration in a predominantly north and south direction. The emphasis is on maximising the amount of occupied space that has access to windows and on minimising the distance from the building core to the perimeter. This can create conflict with programmatic and logistical requirements, such as keeping traffic distances short for staff circulation and materials transportation.

Schematic Design. During the schematic design phase, daylight strategies are about interiors, focusing on spatial considerations to optimise daylight penetration and defining ceiling height, layout, and partition wall transparency with clerestory windows for borrowed light. The planning focus is directed toward coordinating space types that require daylight and views and placing them along the perimeter.

Design Development. During the design development phase, the daylighting strategies' focus is on envelope design to optimise quantity and quality of daylight while minimizing solar gains. The interior design focus is on surface reflectivity and optimising furniture layout to align with visual and thermal comfort requirements.

Construction Documents (CD). Coordination of electrical lighting includes the placement of photo-sensors and occupancy sensors for controlling automated daylight switching and dimmable ballasts.

2.3 Use Daylighting Analysis Tools to Optimise Design

This Guide is designed to help achieve energy savings of 30% without energy modeling, but energy and daylighting modeling programs make

同影响着建筑的造型和轮廓，对建筑骨架的塑形起到了综合驱动作用。

日光策略影响到建筑设计各个阶段的不同层面，总体来讲，日光策略可分为四个方面：

预设计。在预设计阶段，日光策略聚焦于整体研究和楼面的塑形。其目标是使纵深最小化，让空间最大限度地靠近窗户，通过设置光井、垂直井、前厅以及南北朝向的窗户来实现日光照明。设计重点在于最大化靠近窗户的使用空间、最小化楼面中心与外围之间的距离。这可能会造成与项目规划和后勤要求之间的冲突，例如保持员工行走路线和材料运输的最短距离。

方案设计。在方案设计阶段，日光策略侧重于室内，聚焦于优化日光渗透、天花板高度、布局、隔断墙的透明度（可设天窗来借光）等问题的考虑。规划重点在于协调需要日光和观景的空间坐标，将它们设置在建筑外围。

设计开发。在设计开发阶段，日光照明策略聚焦于建筑外壳设计，以优化日光的质量和数量，减少日光吸收。室内设计聚焦于表面反射率和优化家具布局，以满足视觉和热舒适度的需求。

施工设计。协调电气照明，包括设置光电传感器和感应传感器来自动控制日光开关和调光镇流器。

2.3 利用日光照明分析工具来优化设计

本指南旨在帮助建筑设计团队在不用建立能量模型的前提下实现节能30%。但是能量模型和日光照明模型项目

evaluating energy-saving trade-offs faster and daylighting designs far more precise.

Annual savings will have to be calculated with an annual whole-building energy simulation tool after the daylighting design tools have been used to determine the footcandles in the spaces and after the windows have been appropriately sized. Current daylighting analysis tools do not help with heating and cooling loads or other energy uses; they predict only illumination levels and electric lighting use.

2.4 Space Types, Layout, and Daylight

In healthcare facilities, daylight is a key requirement for all occupied spaces. However, the individual lighting needs are different for staff, patients, and the public. The goal is to identify the spaces that best lend themselves to daylight harvesting and saving energy and to recommend layout strategies that allow locating spaces on the perimeter of the building. The potential of energy saving through daylighting varies and depends on programme and space types, which can be broadly characterised by the following four categories of occupied spaces.

Patient Rooms and Recovery Areas. Patient rooms by nature require quality views and daylight. In patient rooms, lighting level requirements are typically low and daylight control is driven by the patient's health condition and individual needs. Prioritising response to patient needs makes the patient room an unreliable space for maximisation of daylight, making it unsuitable as a source for daylight harvesting and energy savings.

Diagnostic and Treatment (D&T) Spaces. Typically dominated by planning criteria, such as circulation distance, proximity, and adjacency requirements, operating rooms and procedure rooms are often located at the core of a deep floor plate with no access to views and daylight. Breaking up the D&T block requires careful planning, but locating these spaces on the building perimeter for daylighting and views is feasible without surrendering flexibility, as case studies have shown (Burpee et al. 2009.; Pradinuk 2008).

Staff Areas (Exam Rooms, Nurse Stations, and Offices). Locating staff spaces

能更快速地衡量节能总量，并且能让日光照明设计更加准确。

年度节能量的计算必须以整体建筑的年度能量模拟工具来进行计算，前提是以日光照明设计工具计算出空间的尺烛光量，且窗户的尺寸也设计得当。当前的日光照明分析工具不能分析采暖和制冷负荷或其他能耗；它们只能预测光照度和电气照明能耗。

2.4 空间类型、布局和日光

在医疗设施中，日光是使用空间的主要需求之一。然而，员工、患者及公共人员区域的照明需求各有不同。设计的目标是明确最适合日光收集和节能的空间，并建议将这些空间围绕着建筑外围设置。各种日光照明策略的节能效果不同，取决于设计规划和空间类型。医疗设施的空间可大致分为以下几种类型。

病房和康复区。病房的特性要求它拥有高质量的景观和日光。在病房，照明度要求相对较低，而日光控制则由患者的健康条件和个人需求所决定。由于病房以患者的需求为首要考虑，它并不是一个可靠的最大化日光的空间，不适合进行日光收集和节能。

诊断和治疗空间。由于受到设计标准——如走道距离、邻近度以及毗邻环境的要求，手术室和处置室通常设在楼面的中心，无法享受风景和日光。打破诊断和治疗空间的阻碍需要进行详细的规划，但是案例研究（波尔皮等，2009；普拉蒂努克，2008）显示，将这些空间设在建筑外围以获得风景和日光不会影响其灵活性。

工作人员区域（检查室、护士站和办公室）。将工作人

on the building perimeter is essential for staff performance, a design strategy that dovetails with the effort to save energy through reduction of electric light and cooling loads.

Public Spaces (Lobbies, Reception, Waiting Areas, and Transitional Spaces). These spaces provide the best opportunity for high ceilings with high, large-scale fenestration and offer the largest potential for daylight harvesting and energy savings due to their depth and potentially high ceilings.

The following recommendations apply to spaces that are not located on the building perimeter but will allow for additional energy savings if they are designed to follow specific rules.

Internal Corridors. In single-storey buildings or on top-level floors, where sidelighting is not available, top lighting should be used to provide daylight for corridors and contiguous spaces. Make sure that nurse stations, which are frequently placed in niches of circulation areas, and waiting areas have access to daylight and views.

Conference Rooms. Conference rooms are densely populated spaces that build up high interior heat loads for only a limited period of time. When located on the perimeter, the interior loads and solar radiation penetrating the perimeter wall accumulate, leading to escalation of peak loads and oversizing of HVAC systems. As a strategy to minimise peak load, conference rooms should be located on north façade perimeters only or inboard, avoiding west-, south-, and east-facing perimeter walls. This approach is supported by prioritising perimeter space for permanently occupied spaces, which make better use of daylight and views than conference rooms, which remain unoccupied in many cases.

From an energy performance standpoint, public areas and staff spaces are the most beneficial spaces for harvesting daylight, which underscores the importance of locating these spaces on the perimeter, preferably in a north- and south-facing configuration. Although patient rooms in hospitals typically have large fenestration and occupy a significant part of the building perimeter, they can't be considered effective sources for energy savings.

员区域设置在建筑外围有利于提高工作效率，还能通过减少电灯和制冷负荷来实现节能。

公共空间（大堂、前台、候诊区和过渡空间）。这些空间适合采用高吊顶和大型高开窗设计，它们的纵深和高吊顶都适合日光收集和节能。

以下建议适合无法设置在建筑外围的空间，但是如果遵循以下规则，仍然可以实现额外的节能。

内部走廊。在单层建筑或建筑的顶楼，如果不能实现侧光照明，则应考虑在走廊及其邻近区域采用上部照明。保证护士站（通常被设计在流通区域的壁龛里）和候诊区可以享受日光和风景。

会议室。会议室的人员十分密集，在短时间内会累积较高的室内热负荷。如果将会议室设在建筑外围，室内热负荷和太阳辐射透过外墙累积，会导致峰值负荷的增加，使空调系统规模过大。为了减少峰值负荷，会议室应当设在建筑北面或内侧，避免与东、西、南朝向的外墙相连。因为会议室在大多数时间都无人使用，这种设计可使外围空间更多地成为永久性使用空间，更充分地利用日光和风景。

从能源绩效的角度来讲，公共区域和工作人员区域是最适合进行日光采集的空间。因此，应当将它们设在建筑外围，最好是采用南北朝向的配置。尽管医院的病房通常拥有大窗户，并且占据了建筑外围的主要部分，它们并不是主要的节能来源。

2.5 Building Orientation and Daylight

Effective daylighting begins with selecting the correct solar orientation of the building and the building's exterior spaces. For most spaces, the vertical façades that provide daylighting should be oriented within 15° of north and south directions. Sidelighted daylighting solutions can also work successfully for other orientations, but they will require a more sophisticated approach to shading solutions, and they would reach beyond the recommendations proposed for accomplishing the goals stipulated in this Guide.

Context and Site. Ensure that apertures are not shaded by adjacent buildings, trees, or by components of the small healthcare facility itself.

2.6 Building Shape and Daylight

Best daylighting results are achieved through limiting the depth of the floor plate and minimising the distance between the exterior wall and any interior space. Narrowing the floor plate will in most cases result in introducing courtyards and articulating the footprint for better daylight penetration.

Building Shape and Self-Shading. Optimising the building shape for daylight translates to balancing the exterior surface exposed to daylight and self-shading the building mass to avoid direct-beam radiation. Effective daylighting requires a maximum amount of occupied area to be located within minimum distance to the building perimeter.

For sunny climates, designs can be evaluated on a sunny day at the summer solar peak. For overcast sky climates, a typical overcast sky day should be used to evaluate the system. Typically, the glazing-to-floor ratio percentage will increase for overcast sky climates. Daylighting can still work for a small healthcare building in a overcast sky climate, however. Overcast sky climates can produce diffuse skies, which create good daylighting conditions and minimise glare and heat gain.

Daylighting systems need to provide the correct lighting levels. To meet the criteria, daylight modelling and simulation may be required. Daylighting systems should be designed to meet the following criteria.

- In a clear sky condition, to provide sufficient daylight, illuminance levels

2.5 建筑朝向和日光

高效的日光照明适于选择正确的建筑朝向以及建筑的外部空间。对大多数空间来说，提供日光照明的垂直立面应当与正南正北的角度不超过15°。侧光照明同样适用于其他朝向，但是它们需要更好的遮阳方案，可能会超出本指南所规定的设计目标。

环境和场地。保证窗口不会被周边的建筑、树木或小型医疗设施本身的其他结构所遮挡。

2.6 建筑造型和日光

限制楼面纵深、缩小外墙与室内空间之间的距离能够实现最佳的日光照明。收窄楼面的同时通常会引入庭院、或将建筑轮廓连接在一起，以实现更好的日光渗透。

建筑造型和自我遮阳。为日光照明而优化建筑造型等同于平衡外墙的日光暴露面积和建筑是防止直接太阳辐射而进行自我遮阳设计。高效的日光照明要求将尽量多的使用区域设置在离建筑外墙尽量近的距离之内。

阳光明媚时，设计可以评估夏季太阳照射峰值时的日光效果。阴天时，应当选用常见的阴天情况来评估该系统。通常来讲，如果阴天天气较多，可以增大玻璃装配面积与楼面面积的比值。然而，即使在阴天，日光照明也可以满足小型医疗建筑的照明要求。阴天会让天空的漫射率增大，创造良好的日光照明条件，并且减少刺眼的光线和热增量。

日光照明系统需要提供恰当的照明度。为了实现这一标准，可以选择日光模型和模拟来进行估算。日光照明系统需要满足以下条件：

* 在晴天，为了提供充足的日光，照明度的范围应在

should achieve a minimum of 25 fc but no more than 250 fc.
• In overcast sky conditions, daylighted spaces should achieve a daylight factor of 2 but no more than 20.

The same criteria for lighting quality and quantity apply to electric lighting and daylighting. When the criteria cannot be met with daylighting, electric lighting will meet the illuminance design criteria. The objectives are to maximise the daylighting and to minimise the electric lighting. To maximise the daylighting without oversizing the fenestration, in-depth analysis may be required.

2.7 Window-to-Wall Ratio (WWR)
There are two steps to approaching window configuration and sizing. The first is that they should follow interior-driven design criteria such as occupancy type and requirements for view, daylight, and outdoor connectivity. The second step targets peak load and energy use, which limit window size to comply with the mechanical systems target. For small healthcare projects to achieve 30% savings, the overall WWR should not exceed 40%.

2.8 Sidelighting: Ceiling and Window Height
For good daylighting in cellular-type spaces, a minimum ceiling height of 9 ft is recommended. In public spaces, which extend to greater depth, such as waiting areas and lobbies, ceiling height, at least partially, should be 10 to 12 feet. When daylighting is provided exclusively through sidelighting, it is important to elevate the ceiling on the perimeter and extend glazing to the ceiling. Additional reflectance to increase lighting levels can be achieved by sloping the ceiling up toward the outside wall.

2.9 Sidelighting: Wall-to-Wall Windows
Raising the window levels to ceiling level is the first priority for deepening daylight penetration. However, to balance light levels in the room and to mitigate contrast, it is equally important to maximise the window width. By extending the window width from wall to wall, the adjacent partitioning walls receive greater exposure and act as indirect sources of daylight while also achieving greater depth of daylight penetration. Even more daylight and a wider range of view can be gained by making the first 2 to 3 ft of the cellular partitioning walls, where they meet the perimeter wall, transparent.

25fc~250fc之间（fc为footcandle，尺烛光的缩写）。
* 在阴天，日光照明空间的采光系数应在2~20之间。

日光照明与电气照明采用相同的照明质量和数量要求。当日光照明无法达到标准时，将由电气照明来实现。设计目标是最大化日光照明、最小化电气照明。为了在不扩大开窗面积的前提下实现日光照明最大化，必须对建筑的纵深进行分析。

2.7 窗墙比
进行窗户配置和尺寸设计有两个步骤。第一步是确定窗户必须满足室内设计要求，例如使用类型、景观要求、日光要求及与室外的连接性等。第二步指向峰值负荷和能源消耗，使窗户尺寸与机械系统目标相一致。对小型医疗项目来讲，要实现30%的节能，整体窗墙比不能超过40%。

2.8 侧光照明：天花板和窗户高度
为了在多隔断空间内实现良好的日光照明，天花板的高度至少应为9英尺（约2.74米）。由于候诊区和大厅等公共空间的纵深较深，天花板的高度（至少是局部高度）应达到10~12英尺（约3.05~3.66米）。如果只有侧光照明时，必须提升靠近建筑边缘的天花板高度并且将窗户延伸到天花板。使天花板朝着外墙向上倾斜可以提高反射率，从而增加照明度。

2.9 侧光照明：整墙窗户
深化日光渗透率的首选是将窗高提升到天花板。然而，为了平衡房间内的光照度、减轻对比，增加窗户宽度同等重要。如果将窗户宽度延伸到覆盖整面墙壁，相邻的隔断墙则会获得更多的光照，从而成为间接的日光光源，还会实现日光渗透的加深。如果将隔断墙靠近外墙的2~3英尺（约0.61~0.91米）设定为透明的，就能够实现更好的日光照明和更广阔的视野。

Nursing Homes
疗养院

Nursing Home Vivaldi

维瓦尔第疗养院

Architects: Seed Architects
Location: Zoetermeer, The Netherlands
Building Area: 12,700m²
Project year: 2010
Photographs:
© Courtesy of Seed Architects
建筑师：希德建筑师事务所
项目地点：荷兰，祖特尔梅尔
建筑面积：12700平方米
完成时间：2010年
摄影师：©希德建筑师事务所

The main challenge for the design of the new nursing home was to make it suitable for the uncertainty of future needs. Due to its layout the existing 25 years old nursing home could not be modified to comply with the current requirements and had to be demolished. Therefore the client Vierstroom, still feeling the impact of writing off the building 15 years too early, was very eager to invest in a sustainable new building.

Through a thorough analysis the architects realised that not the quality of nursing, but the quality of living was the most important factor to achieve the best change of using the building for an set economic lifespan of 40 years or even longer. So the focus of the project shifted from nursing to living.

The layout of the building and its structure includes the possibility to change every wing of the building from nursing into caring. In this way it can be adapted to future needs of care. But even when nursing in the future completely differs to nursing now the building can, when desired, be transformed into a residential building with several

1

1. Water side view of exterior
2. Main entrance with canopy and bicycle parking
3. Detail of window
1. 临水建筑外景
2. 主入口配遮篷和自行车停放处
3. 窗体细部

different types of dwelling. There is an increasing "market" for independently living elderly in a sort of community which needs to be addressed.

With a single corridor situated next to the facade the nursing wings of the building could be column free with floors spanning from facade wall to facade wall. This not only achieves a great value in the future but a very pleasant ambient experience as well. A lot of natural light enters the corridor which are usually cantered in the wings and therefore dark and closed. The passage through the corridor with windows of different sizes on different heights gives a variety of views into the gardens. Moving through the building is therefore never dull both for the (disabled) inhabitants as for the personnel as well.

The layout of the communes which are set up for groups of 8 elderly people offers the opportunity to organize nursing more like a hotel. Four groups can be mixed in a way that people with the same way of living can easily find each other. The positioning of the living rooms/lounges support this freedom of living and making choices. With the current and more individual raised people of the future in mind this might be a very important issue.

The situation on an island surrounded by water creates an urban quality most people want to live in. The water created a smooth kind of privacy to the neighbours without losing spatial quality. The wings of the building are orientated in such a way so that they relate to the directions of the existing surrounding buildings. It not only makes the building a part of the urban fabric but also differentiates the building slightly which makes it interesting to

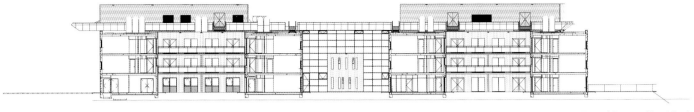

Typical Section (East-West)
标准断面图（东西方向）

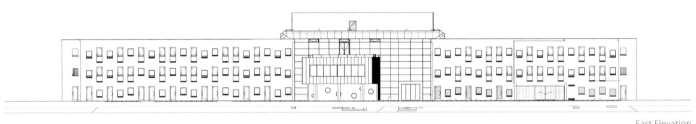

East Elevation
东侧立面图

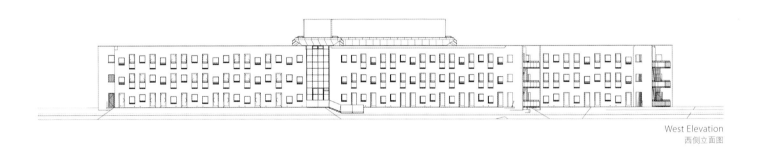

West Elevation
西侧立面图

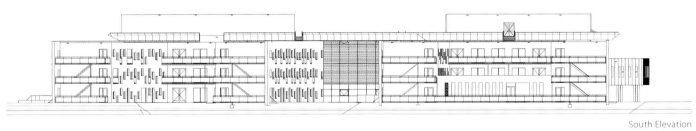

South Elevation
南侧立面图

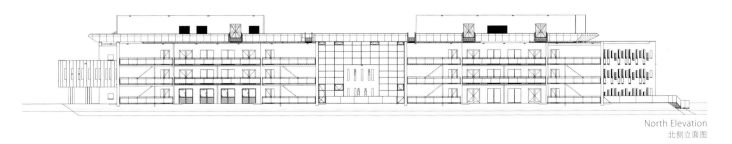

North Elevation
北侧立面图

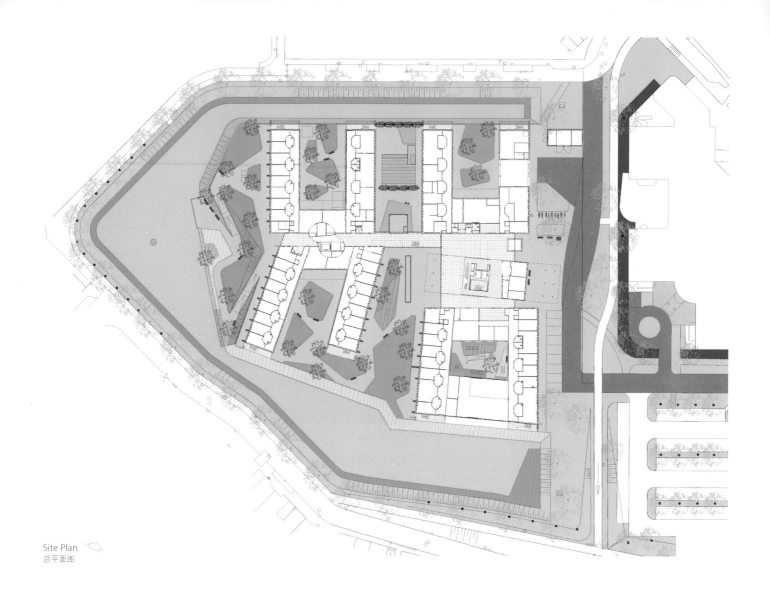

Site Plan
总平面图

move through. Orientation has become far more easy even for demented people.

The use of materials and colours has been very basic. White plaster, wood coloured fibre cement facade cladding, glass and a copper coloured box defines the exterior. The very notable metal box marks the entrance and guides one through the building to the restaurant and terrace at the other side of the lively plaza. The box itself houses mainly general functions like reception, offices and meeting rooms, but also the main staircases and elevators are situated here. The copper coloured cladding at the entrance continues into the interior and after passing the plaza it reveals itself outside at the terrace. The glazed passage gives views to the public outer spaces outside which are bounded by the white plastered wings. On the contrary the courts bounded by the wood coloured cladding are more dedicated to the residents and are therefore more communal. This one finds this difference in atmosphere also in the interior: the plaza and corridor are public while the wings have communal spaces for the group of residents ending in a private room.

The colouring of the interior is related to the outer world. Big photographs of plants, shrubs and trees are used to show differing atmospheres and enhances the relationship with the gardens. Each group of photos is associated to the three different levels of the building: ground, first and second. The extraordinary gardens, the surrounding water, the entrance with a unique plaza, the architectural detailing both in terms of form and used materials make the building very accessible for residents and visitors and ensure that the residents, despite the fact that they are living on an island, keep regular contact with the direct vicinity.

4. Glass curtain wall provide natural light for corridor
4. 玻璃幕墙为室内走廊提供自然光照

5. Courtyard with lounge area
6. Entrance lobby with soft lighting and lounge area
5. 庭院配有休息区
6. 入口大厅灯光柔和，配有休息区

在新维瓦尔第疗养院的设计工作中，最大挑战是适应未来可能出现的需求。当地原有的疗养院已经建成25年之久，由于无法根据当前的需求进行调整，不得不比预计使用年限提前15年拆除。这使得委托方Vierstroom公司迫切地希望能够建成一座可持续发展的疗养建筑。

经过透彻的分析，设计师发现相对于护理质量而言，疗养环境对决定疗养院作为商业建筑的使用年限有着更直接的影响。因此设计的重点也从护理质量转移到了生活环境。

新疗养院的布局和结构使其具备疗养转看护的可能性，以适应未来护理需求可能出现的变化。即使未来的疗养内容与今天完全不同，新疗养院仍旧可以根据需要转化为提供不同居住模式的住宅。有关独居老人的"市场"将在未来逐步扩大。

利用靠近外墙的走廊设计，疗养院的水平跨度中避免了立柱的出现。这样的结构不仅具备极大的未来价值，而且也营造出舒适的氛围。由于放弃了传统布局，走廊里采光充足。花园的美景通过走廊上高度各异、大大小小的窗口映入眼帘，因而无论对住户还是工作人员来说，在楼中穿梭从来都不会是一件乏味无趣的事。

8人为一组的共享空间理念使疗养院的管理工作得以更有效地进行。每4组间的搭配组合方便住户寻找志同道合的伙伴。卧室/休息室的设计为住户提供更多选择的权力。这是对疗养院的未来受众，当下习惯独立生活的年轻人的一种考量。

由于疗养院位于小岛之上，四面环水的优越都市环境具备极大的吸引力，这既为住户提供隐私性又保证了较高的空间质量。建筑的各个部分都通过朝向设计与周围建立了良好的互动，在融入环境的同时保持了自身的独特风味。这样的设计方便人们辨别方向，即使对患有老年痴呆症的住户来说也会容易不少。

工程选用了一些非常基础的建筑材料。外墙使用白色石膏、木质彩色纤维板覆层、玻璃，还有一个铜色盒状结构。这一结构位于入口处，十分显眼。它穿过大楼，一直延伸到广场一侧的餐厅和露

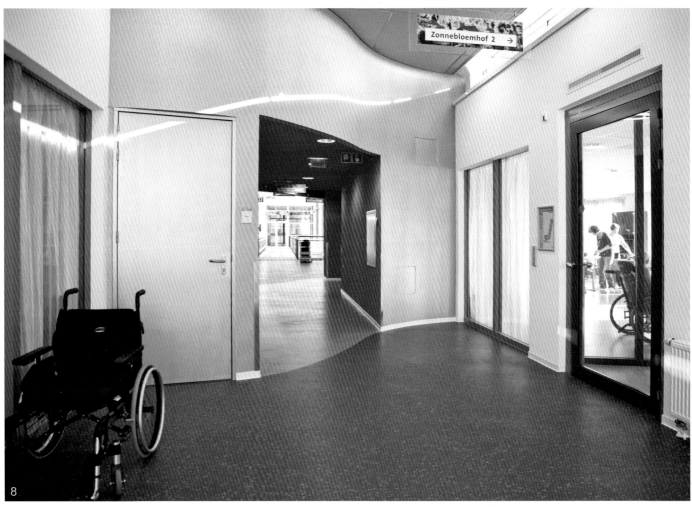

7. Main passage
8. Corridor detail
9. Physiotherapy gym
10. Opening on different level
11. View of corridor
12. Bathroom
7. 主通道
8. 走廊细部
9. 物理疗法健身室
10. 不同高度的开窗
11. 走廊内景
12. 盥洗室

台。盒状结构的主要设施为接待处、办公室和会议室，也包括主楼梯和电梯间。入口处的铜色覆层元素在室内延续，穿过广场之后到达露台和室外。透过玻璃通道可以看到由白色灰泥外墙围出的公共室外活动空间。与此形成对比的是由木质材料包围的院子，这里的氛围更加亲切，因此使用率更高。室内环境也存在这种氛围上的区别。

室内空间的配色与室外风格一致。大幅的植物、灌木和树木照片与窗外的花园呼应，每组照片都与大楼的三个楼层相关。

维瓦尔第疗养院有美丽的花园，怡人的流水，独特的入口，精巧的设计，不仅方便工作人员和住户出入，也为住户提供良好的疗养生活环境。

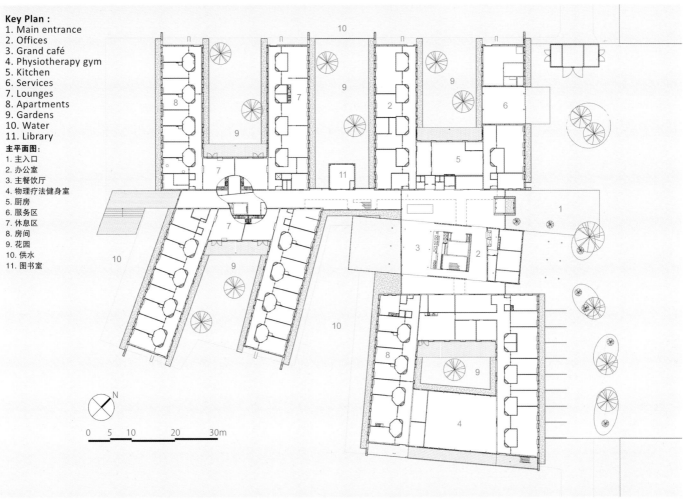

Key Plan:
1. Main entrance
2. Offices
3. Grand café
4. Physiotherapy gym
5. Kitchen
6. Services
7. Lounges
8. Apartments
9. Gardens
10. Water
11. Library

主平面图：
1. 主入口
2. 办公室
3. 主餐饮厅
4. 物理疗法健身室
5. 厨房
6. 服务区
7. 休息区
8. 房间
9. 花园
10. 供水
11. 图书室

Asilo De Ancianos in Baños De Montemayor
蒙特马约尔老年疗养中心

Architects: GEA Arquitectos
Location:
Baños de Montemayor (Cáceres, España),Spain
Gross Floor Area: 3,652.60m²
Project year: 2010
Photographs:
© Ignacio Marqués_PRIMEROS PLANOS
建筑师：GEA建筑师事务所
项目地点：西班牙，巴尼奥斯蒙特马约尔
建筑面积：3652.60平方米
完成时间：2010年
摄影师：©伊格纳西奥·马尔克斯

The project is located in a small town in northern Extremadura called "Baños de Montemayor", surrounded by mountains and forests of chestnut and oak trees. The parcel is in between the urban core (composed of small volumes) and the area of widening of the 1980´s (large buildings). The architects decided to work on a very fragmented structure and volume looking for a smooth transition. From the top of the mountain, the building offers a clear dialogue between the two scales.

The materials are the same ones used on the nearby buildings: granite and white plaster. Granite is chosen as material for contacting the ground, looking for its original texture (bucking granite), so that the pure white volumes would emerge cleanly.

The open public spaces (sometimes half opened), separate day and night zones of the building. The sun penetrates over the built volume that consciously is cut where necessary. Everyday life is made there, in those areas stressed that ultimately make up a core of the lot. The surrounding volumes contain some necessary uses: on one side dining, fitness, healthcaring, nursing, cafeteria ... in the other one sleeps. The bedrooms are all oriented to the west, where the main views are. The corridors

3

are against the mountain and lit by small openings tangent to the path. On the facade of the room's side, the arrangement of recessed windows allows shelter from the midday sun.

As a constructive approach, each bedroom´s toilet is entirely prefabricated and laid on construction site once the floors are raised, saving costs and time. These concrete cells emerge into the corridors qualifying the access to each bedroom, their color differ on every floor in order to avoid confusion for residents.

1. Overall view from the roadside
2. Side view of façade from entrance approach
3. Courtyard with landscaping
4. Atrium with lounge/meeting area
1. 从路边看建筑外观全景
2. 从入口通道看建筑侧景
3. 庭院进行了景观美化
4. 中庭配有休息区和会谈区

Longitudinal Interior Elevation
纵向内部立面图

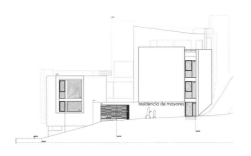

Cross Front Elevation
横向正面立面图

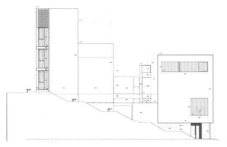

Cross Back Elevation
横向背面立面图

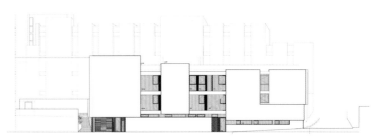

Longitudinal Front Elevation
纵向正面立面图

项目位于埃斯特雷马杜拉北部一个名叫"巴尼奥斯蒙特马约尔"的小镇。这里被群山围绕，生长着茂密的板栗树和橡树。建筑工程在城镇中心（小型建筑）和20世纪80年代扩建的新区（大型建筑）之间进行。

设计师选择通过分散的结构和外形实现一种流畅的过渡。由于工程在山顶上进行，这两个层面都得到了充分的释放和表达。

工程选用了与周围建筑相近的建筑材料，主要包括花岗岩和白色石膏。花岗岩是与地相关的元素，经过与地面原始纹理的对比，呈现纯净的白色质感。

公共开放空间（有时为半开放）将建筑的日间和夜间活动区分隔开来。建筑设计充分考虑到采光的问题，中央为公共活动区，其他功能分别安排在四周：一侧包括餐厅、健身房、保健室、护理室、食堂等，另一侧是卧室。

卧室均为西朝向，视野良好。走廊靠近山区一侧，通过小窗采光。卧室一侧的外墙上，窗口向内凹进，遮挡午间直射的日光。

所有卧室的卫生间都是提前预制的，在完成室内地板后直接安装。这种建设手段既经济又省时。卧室采用混凝土结构，而且每层采用不同的彩色，方便住户辨识。

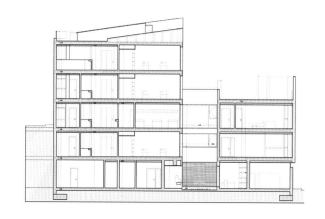
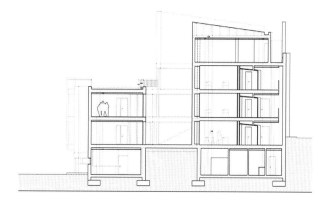
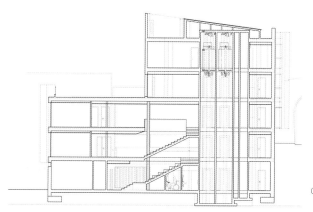

5. Atrium with seats
6-7. Perspective view of corridor
5. 中庭配有座椅
6、7. 走廊透视图

Cross Sections
横截面图

8. Staircase
9. Detail of interior
8. 楼梯
9. 室内细部

Ground Floor Plan
首层平面图

Roof Plan
屋顶平面图

069

Nursing Home Hainburg, Austria
奥地利海恩堡养老院

Architects:
Christian Kronaus + Erhard An-He Kinzelbach
Location: Hainburg, Austria
Building Area: 3,821m² (new building); 4,288m² (old building)
Photographs: © Thomas Ott
Award: Lower-Austrian Building Award 2009, 2nd prize

建筑师：克里斯蒂安·科罗努斯+艾哈德·汀佐巴驰建筑事务所
项目地点：奥地利，海恩堡
建筑面积：3821平方米（新楼）；4288平方米（旧楼）
摄影师：©托马斯·奥特
所获奖项：2009下奥地利州建筑大奖二等奖

In an aging society and due to the tendency in the industrialised west of extended lifetimes, the typology of nursing and retiree homes and its architectural manifestations increasingly gain importance. This happens especially in regards to these buildings' relevance as people's last home in life. In Austria, the public sector, in this case the state of Lower-Austria as the client for this nursing home project, accepts this challenge.

The existing old building of the nursing home in Hainburg was built in 1825 as a castle and changed its use in the course of history for several times: after its original use as a castle it served as an institute for army officers, as barracks, a Russian military hospital and from 1948-1989 as the main hospital of the entire region. Since 2000 it has served as a privately-run nursing home. Recently, the state purchased the building and planned to extend it with a new building housing 50 additional single rooms including their shared facilities.

Conception and Form

The new building's volume – a double-storey, compact bar – is positioned perpendicularly to the existing historic

building. This has a two-fold effect: on the one hand, it only touches the old building punctually through a glazed joint, on the other hand it frames the exterior front yard while preserving a maximum of the park and trees in the back.

Form and structure of the building aim at reconciling the repetitive nature of the program, the wood structure and the need for a fast and efficient construction with the inhabitants' desire for individual differentiation.

Repetition and Differentiation

In order to deal with the desire for individual differentiation, the project employs a strategy of a two-way fold, both on the exterior and the interior.

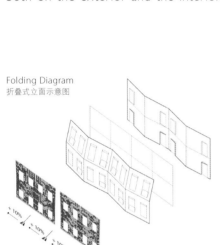

Folding Diagram
折叠式立面示意图

On the outside, the folding breaks the length of the bar volume and allows for local differentiation by making each single-room readable. Each room has 2 windows, one is a regular one and the other is fixed-glazed and tilted, with a low parapet. The latter allows persons that sit in the wheelchair or rest in bed to have an equivalent view out of the window. The regular window is vertical and operable. The resulting niche in the facade that it produces on the outside with its integrated pot can be used for flower planting. Hence, each inhabitant is able to customise his/her face to the exterior. Furthermore, the different tilted windows produce difference through their alternating reflections of the surrounding trees and the sky.

On the inside, the strategy of the fold is employed in the corridor walls. The folding marks each individual single room and generates niches in front of the entrance doors that can be occupied by the inhabitants in various ways, thus rendering the corridor as more than just a space for circulation. Furthermore, the folds shorten the length of the corridor visually and they zone it. Inside the corridor walls, the folds create space for integrated mechanical shafts and built-in closets alike. The latter contain nursing material that is accessible from outside the rooms.

The corridor has 2 common public spaces at its ends. A recreational room is located towards east with views to the close Slovak capital Bratislava. On the west, there is an open loggia with views to Hainburg castle. In the center, in close proximity to the central nurse's ward located at the joint between the old and new buildings, there is a third common room facing south. All three common spaces have balconies/terraces attached. It is this distribution of the common shared spaces that complements the individual privacy of the single-rooms and supports social gatherings in small groups among the elderly at every time of the day.

Adaptive Colour Gradient

In addition to the strategy of the fold, the skin employs a second performative framework to break monotony and

1. Main view from north
2. The old and the new in juxtaposition
3. Façade performance — global coherence and local differentiation through 3-dimensional folding and pixelated colour grading

1. 从北侧看建筑
2. 新旧建筑毗邻而立
3. 外立面特性——全球统一性与地方差异性通过外立面三维折叠和像素色彩分级实现

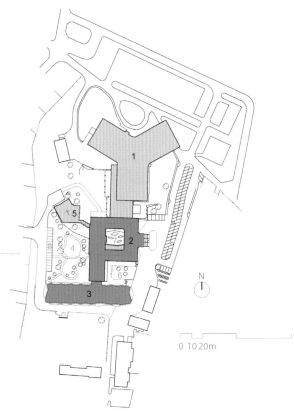

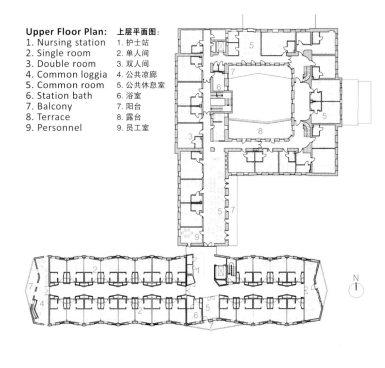

Upper Floor Plan: 上层平面图：
1. Nursing station   1. 护士站
2. Single room   2. 单人间
3. Double room   3. 双人间
4. Common loggia   4. 公共凉廊
5. Common room   5. 公共休息室
6. Station bath   6. 浴室
7. Balcony   7. 阳台
8. Terrace   8. 露台
9. Personnel   9. 员工室

Site Plan: 总平面图：
1. Hospital   1. 医院
2. Existing nursing home   2. 原有疗养院
3. New nursing home   3. 新疗养院
4. Park   4. 停车场
5. Chapel   5. 小礼拜堂
6. Café courtyard   6. 餐饮庭院
7. Therapy garden   7. 治疗花园
8. Parking   8. 停车场

Ground Floor Plan:
1. Café courtyard
2. Main entrance
3. Therapy garden
4. Park
5. Nursing station
6. Single room
7. Double room
8. Common loggia
9. Common room
10. Station bath
11. Café
12. Terrace
13. Seminar room
14. Office

首层平面图：
1. 餐饮庭院
2. 主入口
3. 治疗花园
4. 停车场
5. 护士站
6. 单人间
7. 双人间
8. 公共凉廊
9. 公共休息室
10. 浴室
11. 餐饮区
12. 露台
13. 研究室
14. 办公室

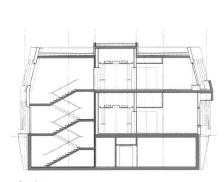

Section
剖面图

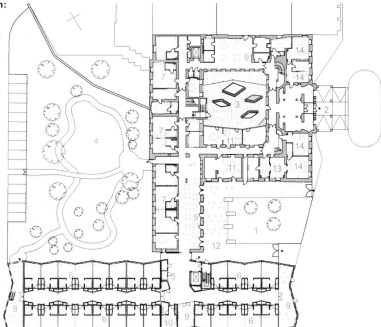

073

Diagram of Colour Gradation
色彩分级示意图

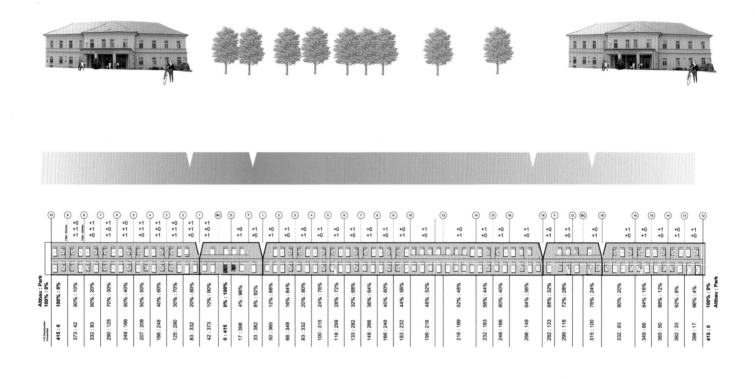

instill differentiation, while maintaining overall coherence. The skin, not unlike a chamaeleon, mediates between two predominant colours: the pink color of the old building and the green of the recreational park in the back. It gradually changes from one to the other, starting with pink at the eastern end of the joint between old and new, changing to green and changing to pink again when returning to the joint on the western side.

The skin is pixelated through the use of diamond-shaped Eternit-shingles. A simple algorithm is introduced, based on a combination of successive mirroring at the fold lines and the gradual addition of pixels of one colour while subtracting the other. This set of basic rules allows for the described performance. At the same time, it keeps the manual and systematic installation by the construction workers simple.

Construction and Finishes

In order to satisfy the need for standardization in order to realise the complex geometry cost-efficient and in a short amount of time, the building was built in timber panel construction with solid wooden floorplates. The advantages comprise the dry mortarless construction with a low deadload and the high amount of prefabrication with its benefits of precision and speed. The single polygons of the walls that result from the folding, were prefabricated piece by piece. The plumbing units were premanufactured as fully-installed, light-construction boxes that only had to be plugged in on-site.

In terms of finishes, the exterior consists of diamond-shaped Eternit-shingles in 2 colours, while the interior is composed out of wood surfaces, laminate and linoleum and wooden floors.

Material and Sustainability

The building fulfills low-energy standard. In addition to the sustainability of the buildings low energy consumption, a few ecological and sustainable materials were consciously applied: the entire structure is made of wood, no reinforced concrete is used aboveground. Only the lift shaft is not built of wood but fire-rated steel, due to fire safety reasons. The thermal insulation is achieved through natural flax, and the gravel bed under the foundations consists of foamed glass made of recycled glass.

4. Main corridor view
5. Corridor performance – breaking the length, defining room entries and integrating storage space through 2-dimensional folding
4. 主走廊
5. 走廊特性——突破长度，确定房间入口，通过二维折叠处理将存储空间融入其中

6

随着西方发达国家出现老龄化社会，人均寿命延长的趋势，疗养院和敬老院的类型调整以及它们在建筑上的体现变得越来越重要。考虑到养老院对一些人来说是生命中的最后一站，这些调整和设计显得尤为意义重大。在这样的情况下，奥地利的公共部门，即下奥地利州政府接受了这个具有挑战的项目。

海恩堡养老院的原有建筑始建于1825年，最初是一个城堡，随着朝代更替，其用途也多次发生改变。先后做过陆军基地、军营、俄国军事医院。直到1948～1989年，它才成为当地的主要医院。2000年起变成私营养老院。下奥地利政府最近将它买下，计划将原有结构扩展成能多容纳50个单人房间和公共设施的新建筑。

概念和形式

新楼为双层竖向结构，矗立在原有建筑旁边。这种布局有两点好处：只通过玻璃结构与原有建筑相连；另一方面新楼将前院围起，同时最大程度地保留了建筑后方的公园和树木。

新建筑的形式和结构考虑到了工程重复的特点，木质主结构和对高效施工的需要，以满足住户对私人空间的要求。

重复和变化

为了满足住户对私人空间的需要，工程采用了双向折叠的室外室内设计。建筑外部的折叠效果在水平层面上打破建筑的竖直感，使每个房间都清晰分明。每个房间有两个窗户，一个形状规则，另一个为有矮墙围绕的固定倾斜玻璃窗。后者方便住户坐在窗前或者躺在床上欣赏窗外的风景。规则的窗口可以在垂直方向打开，而它在外墙上形成的凹陷可以用于花卉种植。每个住户都可以在窗口获得一个独特的视角。树木和天空映照在不同角度的斜窗上，呈现各不相同的风景。

室内的折叠设计体现在走廊墙壁上。每处折叠都是一个房间，在入口大门处形成的凹陷可以服务多种功能。这种折叠设计在视觉上缩短了走廊的长度，并将其分区化。折叠设计使走廊墙里有空间安装综合机械轴和内置壁橱。后者容纳的护理材料也可以从屋外获取。

走廊尽头有两个公共活动空间。其中一个活动室朝向东方，遥望斯洛伐克首都布拉迪斯拉发。西侧的开放长廊则俯瞰海恩堡古城。建筑中央新旧楼交界处，靠近看护病房的地方是另一个南向的公共活动室。这三个活动室都有相连的阳台/露台。这样的公共空间设计将单人房间紧紧整合在一起，鼓励老人在不同时段进行小型聚会活动。

渐变色彩

除了折叠的设计，外墙还采用了第二层装饰性框架，在保持整体感的前提下打破单调，制造渐变效果。外墙的效果就像变色龙的皮肤，由两种主要色彩过渡而成：旧楼的粉色和楼后休闲公园的绿色。墙体颜色逐渐过渡，新旧楼在东端的交界处是粉色的，向外逐渐变成绿色，然后又在西侧连接处变回粉色。

外墙的变色效果是通过菱形艾特尼特单元实现的。除此之外，工程还使用了基于对折叠线连续反射简单的方法，逐渐增加反射色彩的同时减少另一种色彩。这组设计不仅能够实现渐变的效果，还能保证建筑工人进行手工和系统安装工作的简单快捷。

建筑和装饰

为了满足建筑标准化的需要，高效省时地完成复杂的几何建筑工程，项目选用木制面板和实木地板。其优点是干灰泥具有较低的静载量，而大量预制元素的特点则是精度高，速度快。多边形墙壁是分块预制的。水管也是按单元预先安装好的，只需在现场连接即可。

建筑外墙由双色菱形艾特尼特装饰，室内则选用木质建材，油毡压层板和木质地板。

建筑材料和可持续性

项目设计遵循低能环保标准。除了低能耗的环保设计，项目还使用了一些生态材料和可持续性材料。整个结构使用的是木头，地面以上没有使用钢筋混凝土。出于防火安全的考虑，电梯井是唯一使用防火钢而非木头的部分。保温层使用的是天然亚麻，地基下方的砾石层由回收玻璃制成的泡沫玻璃构成。

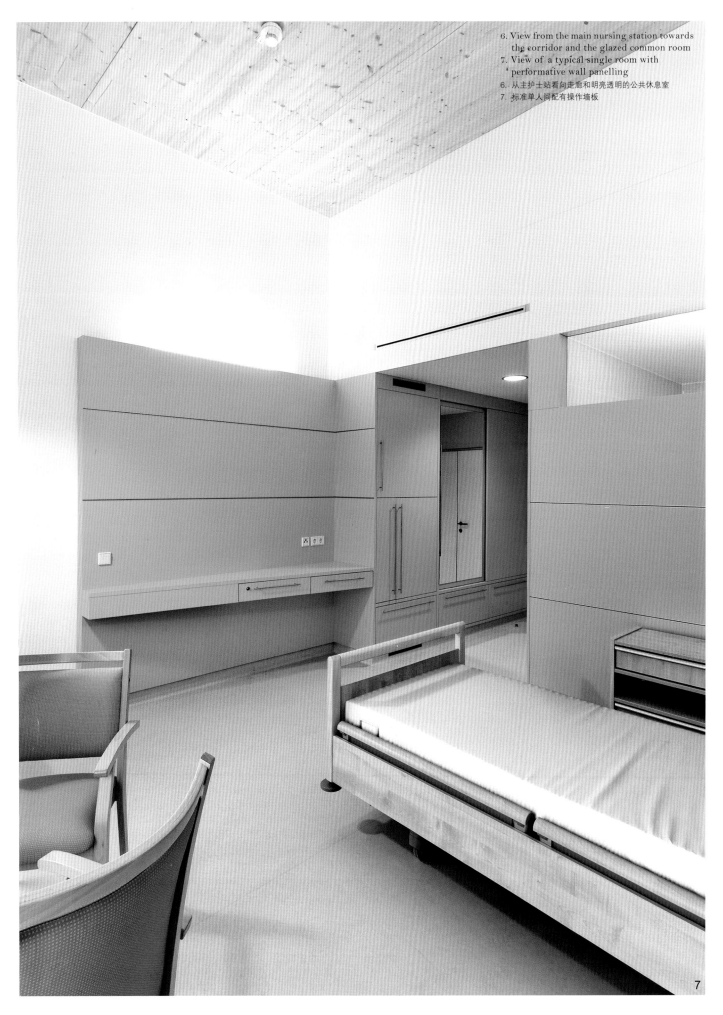

6. View from the main nursing station towards the corridor and the glazed common room
7. View of a typical single room with performative wall panelling
6. 从主护士站看向走廊和明亮透明的公共休息室
7. 标准单人间配有操作墙板

Pflegeheim St. Lambrecht
圣兰布雷希特临终关怀中心

Architects:
Gerhard Mitterberger
Location:
Austria
Project year:
2010
Photographs:
© Courtesy of
Gerhard Mitterberger
建筑师：格哈德·米特伯格建筑事务所
项目地点：奥地利
完成时间：2010年
摄影师：©格哈德·米特伯格建筑事务所

40 nursing beds and infrastructure were largely housed in the new annex, while the restored existing buildings remained the administration, staff rooms, a medical practice in the EC and nursing home beds and cafeteria on the ground floor and public housing, or assisted living in the attic. The project was completed for the community of St Lambrecht (original owner) of the CEU in construction law relationship and is used by the charity as the new operator.

The addition was built under environmentally high standards in timber. The old lake house building continues to be the central part of the nursing home and is exempt from smaller applications such as balconies, canopies angeflickte, etc.

The extension fits flat, one floor, to floor height on the upper floor of the gently rising ground, and thus follows the settlement pattern of the main street of St Lambrecht: Townhouses at the street front and frayed outbuildings, mostly used for agriculture, branching out into the narrow strip of green space behind it.

All rooms are wheelchair friendly on the basement level of the upper floor, which of the annex to the green open space, corridors and common rooms are illuminated by an interior atrium and offers

1. Exterior detail – porch
2. Overall view
3. Courtyard with café
4. Courtyard view from interior corridor
5. Interior lounge

1. 外观细部——门廊
2. 外观全景
3. 庭院配有餐饮区
4. 从室内走廊看向庭院
5. 室内休息区

a protected micro-climate – the natural environment by rinsing the building.
The inner world of the house is indeed for the care, limited mobility residents becoming the only real experienced habitat, and covers intact, thanks to the rural environment that is dominated by the Abbey of St Lambrecht, out there with memories of life.

The chapel was redesigned by Oberwalder Zita, the central themes are awakening (the sunrise on Mount Sinai as a slide on the east wall), and Benedict of Nursia, the founder of the Benedictines, a reference to the Abbey of St. Lambrecht (section of the ceiling fresco of San Benedetto/Subiaco)

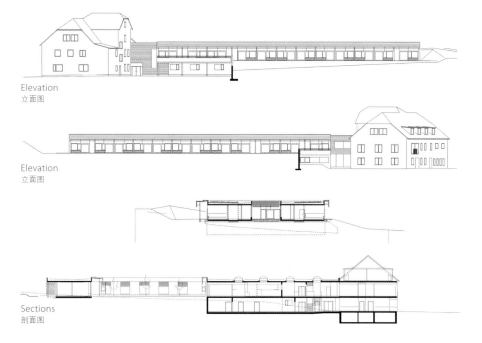

Elevation
立面图

Elevation
立面图

Sections
剖面图

关怀中心的新楼包含40个护理床位和一些基础设施。修整后的原中心大楼的一楼容纳了行政办公室、员工室、医疗室和食堂，顶楼作为辅助生活区使用。

根据相关的建筑法规，新建的关怀中心为圣兰布雷希特社区服务，其价值也将在慈善事务中得到体现。

关怀中心的新楼使用大量木材，严格执行环保标准。除了添加阳台、檐篷等细节，别墅式的旧楼仍作为核心结构使用。扩建部分顺应上升的地势，遵循圣兰布雷希特主要街道的分布规律。临街建筑向狭窄的绿色空间延伸扩展。

高层结构里第一层的所有房间方便轮椅通行，绿色开放空间、走廊和公共活动室通过中庭采光，形成微气候。建筑的室内环境充分考虑到行动不便的住户的需求，浓厚的乡村氛围使人联想起昔日的美好回忆。

小礼拜堂由欧泊瓦尔德·兹塔重新设计。主题是"苏醒（东墙上是西奈山日出的景象）"和本笃会创始人努尔西亚本笃，向圣兰布雷希特修道院致意（圣兰布雷希特/撒贝卡天花板壁画的节选）。

6. Chapel
7. Interior detail
6. 小礼拜堂
7. 室内细部

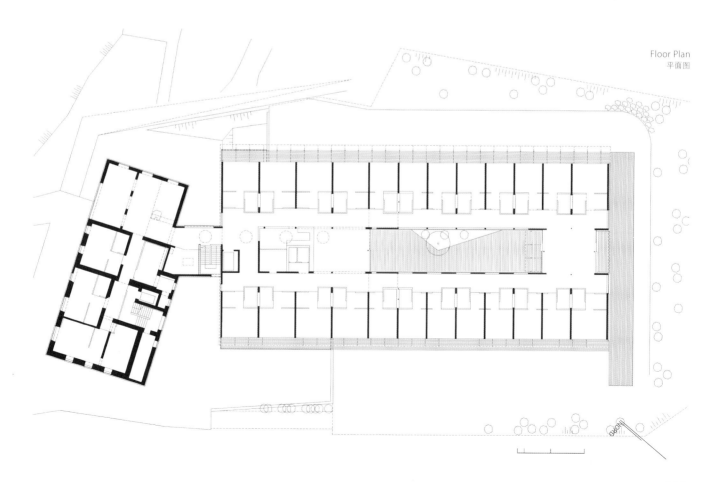

Floor Plan
平面图

Renovation and Enlargement of a Health Centre in Seregno (MB)

塞雷尼奥卫生中心改造扩建工程

Architects: Mondaini Roscani Architetti Associati
Location: Italy
Building Area:
4,000m² (1st phase);
912m² (2nd phase)
Project year: 2012
Photographs: © Marco Capannini, Archivio Mondaini Roscani
建筑师：
蒙达伊尼·罗斯卡尼联合建筑事务所
项目地点：意大利
建筑面积：一期工程4000平方米；
二期工程912平方米
完成时间：2012年
摄影师：©马克·卡帕尼尼、
蒙达伊尼·罗斯卡尼资料库

"Regeneration as Paradigm of Future"

The enlargement of the original building of Ronzoni's Foundation & Villa belongs to a wider project, which won the "EUROPAN: european competition for new architectures". The project is located in an expansion area in the city of Seregno, composed of various building types and open spaces, which determine a general lack of identity. The project aims to regenerate the existing buildings, trying to restore that identification between buiding and citizen, that made public Italian architecture so relevant in the past. Therefore the design concept consists in restoring degraded open spaces through volumetric densification.

The historic building was built in 1930. In the 1980s a new rest home was built next to this building becoming soon the main centre, while the older one lost its role and centrality. The project of restructuration and enlargement keeps the main monumental facades, facing the perimetric streets, while the back facade, facing an empty courtyard, is completely demolished. Its reconstruction becomes the occasion to enlarge its gross surface, fixing and laying different volumes upon

1. Back façade viewed from roadside
2. Side view of exterior
3. Entrance and approach for the disabled
4. Façade detail
5. Passage

1. 从路边看建筑的背面
2. 外观侧景
3. 入口及无障碍通道
4. 外立面细部
5. 通道

such wall. These projecting elements contain the new vertical connections and also new facilities. The 1st phase consisted in building new staircases and elevators and elevating one floor, while connecting it to the renovation of the loft. The 2nd phase focused on the enlargement of services and the realisation of new recovery rooms located inside the new volume projecting on the courtyard. The building hosts three different fonctions such as assistance services, ambulatories and gyms for phisiotherapy rehabilitation, which are divided by floors: at the ground floor there are health care services, gyms, ambulatories and dressing rooms for employees; at the 1st floor there are specific services for the development age, play-areas, cafeteria, recovery rooms for children; at the 2nd and 3rd floor there are apartments and accomodations for self-sufficient elder people. (Text by Gianluigi Mondaini)

5

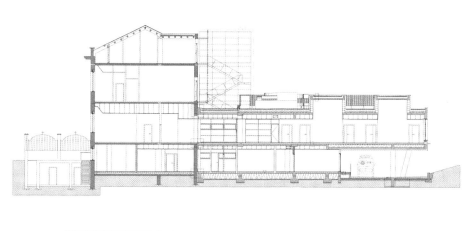

Section
剖面图

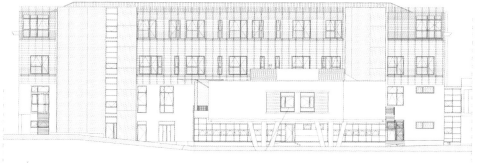

Elevation
立面图

6. Balcony with canopy
7. Corridor view with atrium
6. 带遮篷的阳台
7. 走廊及中庭

未来典范的产生

这项工程是对原来的龙佐尼基金会及别墅的扩建，并在"欧洲新建筑竞赛"中获奖。工程位于塞雷尼奥市的开发区，这里的建筑类型各异，布局松散，缺乏统一的规划感。工程的主要目标是改造原有建筑，重建意大利公共建筑中传统的居民——建筑联系。因此工程的设计理念围绕体积密集下递降的开放空间展开。

卫生中心始建于1930年。20世纪80年代紧挨旧楼而建的副楼很快取而代之，成为了新的业务中心。在这次重建扩建工程中，朝向街道的主外墙保留，正对庭院的后墙完全拆除。借着重建的机会，设计师决定扩大卫生中心的总面积，并加强建筑的空间感。这些想法通过新的垂直结构和新设施得以实现。一期工程包括新楼梯间和电梯的修建，二期工程则以扩大服务区面积和建设新康复区为重点。新卫生中心的三个主要功能是救助服务、门诊和物理疗法馆，分别位于不同楼层。一楼是保健服务中心、活动室、门诊处和职工更衣室；二楼是发展阶段特殊护理区、游戏区、食堂和儿童病房；三楼和四楼是老年公寓。（文／吉安鲁基·蒙达伊尼）

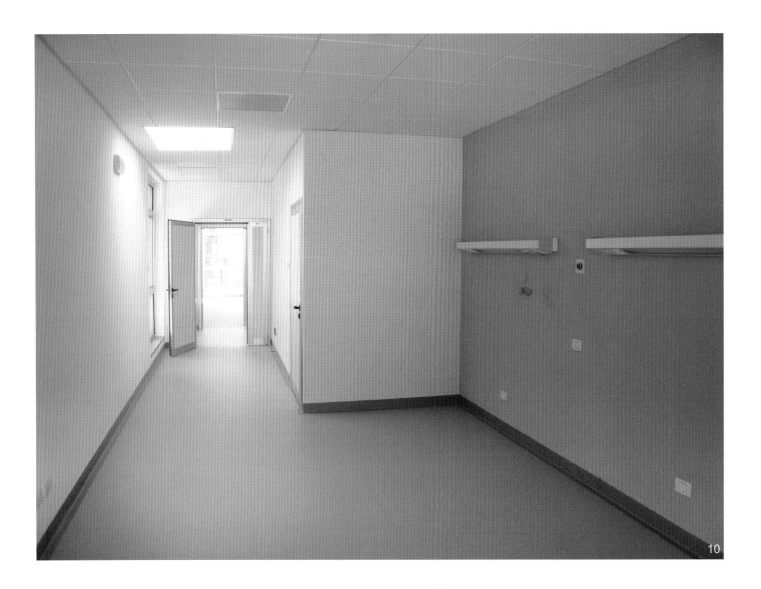

8-9. Perspective view of corridor
10. Interior view of patient room
8、9. 走廊透视图
10. 病房内景

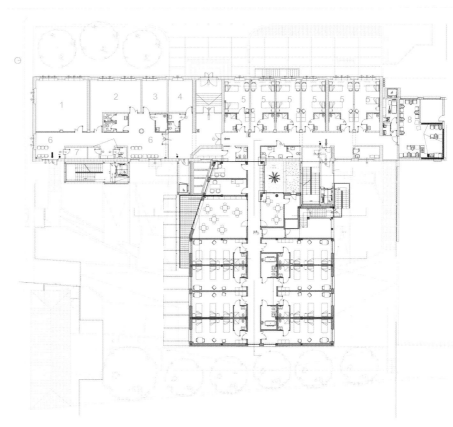

Floor Plan:
1. Sensory stimulation gym room
2. Teach room
3. Manipulations
4. Doctor's office, psychiatric and physiatrist
5. Hospitalisation
6. Waiting
7. Storage
8. Living room with kitchenette

平面图：
1. 感官刺激健身室
2. 教学室
3. 处理室
4. 医生办公室，精神治疗与物理治疗
5. 留院治疗区
6. 等候区
7. 存储空间
8. 起居室配小厨房

Hanzeborg Care Centre
翰兹伯格护理中心

Architects:
Dutch Health Architects
Location: Lelystad, Netherlands
Building Area: 15,568m²
Project year: 2010
Photographs:
© Courtesy of EGM architecten
建筑师：荷兰健康建筑设计事务所
项目地点：荷兰，莱利斯塔德
建筑面积：15568平方米
完成时间：2010年
摄影师：©EGM建筑公司

Hanzeborg care centre is located in Hansa Park, a new residential area adjacent to the city centre of Lelystad. A contemporary yet timeless and unique building where providing and receiving care has a new dimension added to it. EGM used the initial design as the basis for this building. A building where care, services and activities all come together. It is the combination of different disciplines which gives strength to the complex. Hanzeborg has a strong position within the local urban dependency. In the present social context, a private healthcare is no longer restricted to just one location.

However, care should remain easily accessible. Therefore, in the design phase, the relationship between interactive facilities was kept in check. Social and recreational activities ran seamlessly together with the actual care. Hanzeborg had become the beating heart of the neighbourhood, therefore, the two lower levels – where public functions are embedded – were designed as transparent and accessible as possible. For those enjoying a walk from the care facility to the park, located behind the building, interesting pedestrian routes through Hanzeborg are now offered. This building combines its transparent structure and public access to the entrance of the residential area through the park, forming an interesting point on the route.

1

1. Overall view from roadside
2. Side view of exterior from street
1. 从路边看建筑全景
2. 从街边看建筑侧景

汉莎公园是一个靠近莱利斯塔德市中心的新住宅区，翰兹伯格护理中心就坐落在这里。这座现代而又经典的独特建筑在新层面上丰富了护理工作和治疗体验。EGM建筑公司将此作为设计的出发点，努力实现一个将服务与活动结合的建筑。翰兹伯格护理中心在当地的城市布局中占有重要的位置。在当前的社会背景下，私营的护理中心也不再局限于单一的地点。

然而护理设施应该首先保证通行方便，因此在设计阶段，互动设施的相互关系是考虑的一个重点。社交和娱乐设施都与护理设施紧密地联系在一起。翰兹伯格护理中心是当地的核心建筑，所以行使公共职能的较低楼层的设计尽量透明且便于通行。充满趣味的步行路线连接护理中心和汉莎公园，方便人们在户外散步。透明结构和通行路线是护理中心最突出的特点。

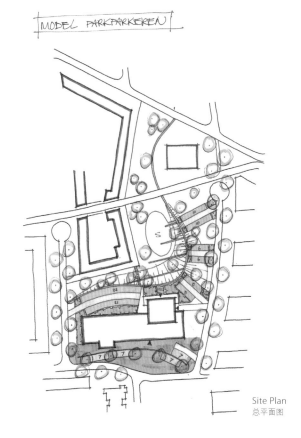

Site Plan
总平面图

Elevations (Upper Four)
立面图 （上方四图）

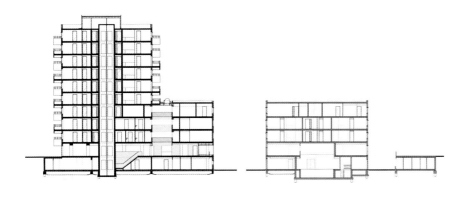
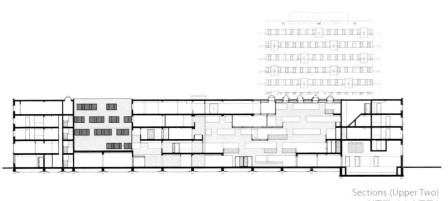

Sections (Upper Two)
剖面图 （上方两图）

3. Entrance lobby and reception
4. Entrance lobby viewed from upper level
3. 入口大厅及接待处
4. 从楼上看入口大厅

Ground Floor Plan:
1. Local centres
2. Office centres
3. Pharmacy
4. Hair salon
5. Wijkpost
6. Laundry & workshop
7. Individual apartments
8. Technique
9. Physio/fitness
10. Pool and changing rooms
11. Entrance
12. Shop
13. Kitchen & grand café centres
14. Hall centre

首层平面图：
1. 当地居民中心
2. 办公中心
3. 药房
4. 美发沙龙
5. 邮寄服务
6. 洗衣店和工作室
7. 单人房间
8. 机械室
9. 理疗/健身
10. 泳池与更衣室
11. 入口
12. 商店
13. 厨房与主餐饮中心
14. 大堂中心

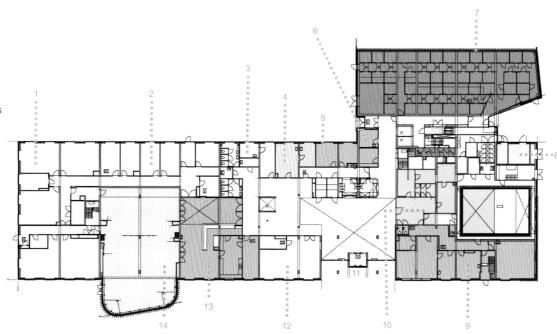

5. Swimming pool
6. Café
7. Lounge/meeting room
5. 游泳池
6. 餐饮区
7. 休息/会谈区

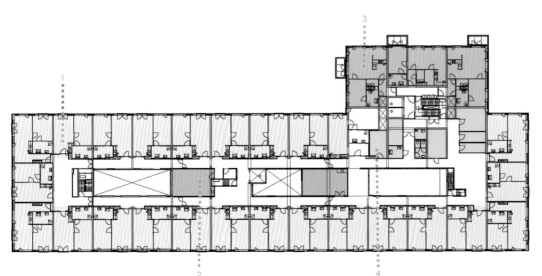

Third Floor Plan:
1. Short stay apartment
2. Multifunction space
3. Three larger apartment (rent)
4. Additional care functions: meeting room, central bathroom & toilets, storage

四层平面图：
1. 短期居住单间
2. 多功能空间
3. 三个较大的单间（出租）
4. 附加功能：会议室、中央浴室和卫生间、储藏室

Pflegeheim Schorndorf
舍恩多夫疗养院

Architects:
ippolito fleitz group / Peter Ippolito, Gunter Fleitz
Location:
Schorndorf, Germany
Building Area: 1,322m²
Photographs:
© Zooey Braun
建筑师：伊波利托-弗莱茨集团/彼得·伊波利托，冈特·弗莱茨
项目地点：德国，舍恩多夫
建筑面积：1322平方米
摄影师：©佐伊·布劳恩

The main objective of the interior design was to create rooms that emphasise the personality and individuality of the nursing home's residents. The design concept promotes their independence, while at the same time responding to the specific needs of elderly people. Even during a stay in a nursing home, its residents should be able to lead a life of dignity and security, and ideally, feel completely at home there.

The home is on a single level that is organised in a clear and structured way. All rooms are wheelchair-accessible and can be easily located in one of the two wings. The spacious recreation areas establish homelike focal points, which support easy orientation.

Thanks to specially-positioned seating niches, the corridors, which would otherwise just be spaces to move through, become communication zones. Each niche is equipped with two comfortable chairs, a small table and a table lamp, and is protected and contained by a tapered, translucent curtain. This creates an island with an almost intimate feel: the perfect retreat for a private discussion or reading a book, while at the same time remaining in contact with the outside world.

1. Perspective view of corridor
2. Lounge area
3. Corridor is also a communication zone
4. Treatment room
5. Larger lounge area
1. 走廊透视图
2. 休息区
3. 走廊也是交流区
4. 治疗室
5. 较大的休息区

Floor Plan
平面图

Recreation areas are tailored towards residents' different needs: privacy in the surroundings of your own room, communication and taking an active role in public life from the comfortable arm chairs in the recreation areas, partial seclusion in the corridor seating niches or active participation in the restaurant, therapy areas or on the terrace in the fresh air.

The materials and colours concept is designed to create a homelike, non-institutional atmosphere, and also to stimulate. Walls are painted in light tones in accordance with a differentiated colour scheme, or decorated with wallpaper. The oak furniture forms a gentle contrast and conveys a sense of home, which is complemented by the warm, greyish-brown linoleum floor. Recreation areas are highlighted by means of inlaid carpet islands.

No two bedrooms are the same. Keeping with the idea of individualisation, each room has its own palette of colours. Wall colour, accent colour and curtains harmonise to create a new combination and different mood for each room. Their form and materiality make standard fittings appear familiar and comforting. And generous shelving and storage possibilities create numerous ways of customising each room with personal touches and effects. All this serves to make each room unique.

The orientation system picks up on this idea. In place of abstract room numbers, each room bears a proper name: "Castle", "Palace" or "Home" – each name referring to some kind of living situation. The rooms are arranged alphabetically to make it easier to find your way around.

A frame is affixed next to the door of each room. This contains the name of the resident and one or two pictures from their past. In this way a relationship between the person and the location is established, creating a bridge connecting inside with outside, yesterday with today.

In addition to the seating islands, the main corridor is characterised by a picture wall that stretches along its entire length. Collective and individual memory fragments from residents' lives are amassed in more than 170 picture frames: record covers from Beethoven to the Rolling Stones, doilies, silhouettes, poems, old maps, postcards, posters, photographs and even whole books. Their sheer variety and number represents the individuality and uniqueness of residents' life stories. To view them is to immerse yourself in the memories of others. You naturally associate your own experiences, smile and exchange stories about what you've seen. At the same time the collection is constantly changing as residents change. Images come and go and there are always new things to discover – the wall keeps on "moving".

The picture wall underscores the design concept: the dignity, individuality and independence of residents is emphasised and communication is fostered so that the institutional character of the nursing home fades almost completely into the background.

6

6-7. Dining room
6、7. 餐厅

舍恩多夫疗养院室内设计的主要目标是创造重视个性与隐私的居住空间。这个设计理念鼓励独立的生活模式，同时也能够满足老年人的特殊需求。即便是住在养老院，老人们的尊严和安全感也应得到最基本的保障，他们在这儿更应该有宾至如归的感觉。

疗养院仅有一层，平面布局清晰合理。所有的房间都适合轮椅通行，且标有清晰的指示。宽敞的休闲区域让住户感受家一般的温馨。

走廊通常只是一个连接性的过渡空间，舍恩多夫疗养院特别的座椅设计为走廊增添了社交活动的功能。每个位置配有两把舒适的座椅、一个小桌和一盏台灯。锥形的半透明窗帘将其围绕，形成保护；这样，既与外界保持一定的联系，又适合进行私人谈话或者细细品味一本好书。

休闲区的设计满足不同住户的多种需要：可以享受私人空间，或坐在休闲区舒适的椅子上与人交谈，参与公共生活；可以在走廊寻得清静，也可以在餐厅积极活动，或是在露台治疗室感受新鲜的空气。

建筑材料和配色的选择都为打造温馨、舒适的居住环境服务。墙壁或涂以浅色，或以壁纸装饰，色调十分特别。橡木家具与之形成微妙的对比，配合棕灰色的油毡地板衬托出家的感觉。镶嵌地毯是休闲区的一大特色。

卧室设计各不相同，每个房间都使用了独特的配色方案，以实现个性化的设计目标。墙色与其他软装元素的色彩协调一致，营造出新颖多变的氛围。这样的形式和搭配使普通的标准元素变得亲切而熟悉。充足的储物空间让每个房间都可以呈现个性化的效果。这些都是强调个性的有效手段。

房间指示方法也延续了这个理念。除了房门号码，每个房间都有一个名字，比如"城堡"、"宫殿"和"家"——每个名字都道出了某种生活状态。房间名称按字母顺序排列，方便查找。

每个房门旁边都有一个小框，标有房间主人的名字，并贴着一两张主人的生活照。这样，人与房间的关系即被确立，在门内与门外，今日与昔日之间构成一个纽带。

除了具有浪漫情怀的座位设计，占了整个墙壁的照片墙是大走廊的另一个特点。170多个相框记录了疗养院住户们共同或各自的记忆：从贝多芬到滚石乐队，各种各样的唱片封面；桌布、剪影、小诗、旧地图、明信片、海报、照片，甚至整本的书。住户的人生旅程都体现在了这些品种繁多的藏品上，照片墙的内容也会随着住户的变化而发生改变。照片出现又消失，照片墙上总有新内容等待细心人发现。

疗养院的设计理念在照片墙上得到了强化：住户的尊严、个性和独立得到尊重和欣赏，人与人之间实现了沟通，养老院式的环境也彻底消失不见。

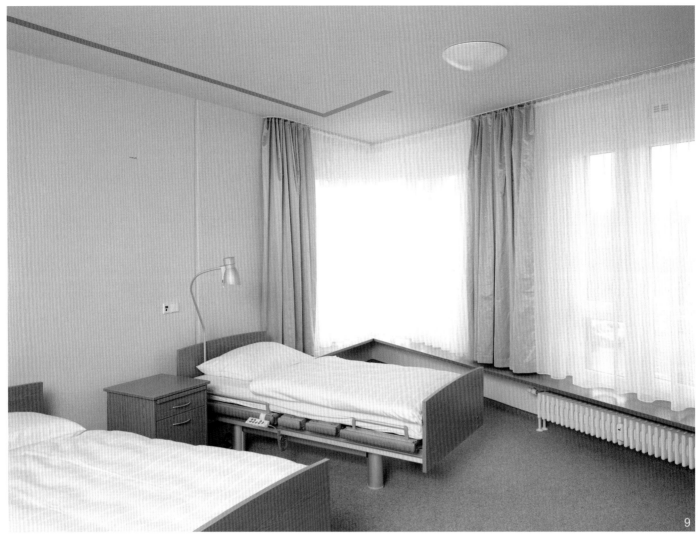

8. Reception/service
9. Two-bed room
10. Detail of bedroom
8. 接待/服务
9. 双人间
10. 卧室细部

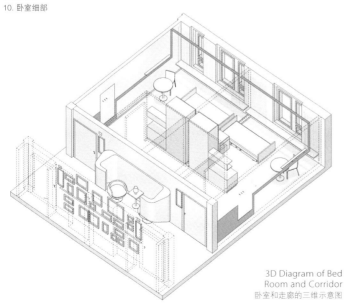

3D Diagram of Bed Room and Corridor
卧室和走廊的三维示意图

Specialised Clinic in Addiction Treatment Hoechsten, Bad Saulgau

巴德绍尔高-赫斯特恩戒瘾治疗中心

Architects: weinbrenner.single.arabzadeh
Location: Nürtingen, Germany
Site Area: 24,600m²
Construction Area: 3,860m²
Project year: 2010
Photographs: © Gerd Jütten - Fotodesign
Award: AIT Award 2012; Global Award for the very best in Interior and Architecture; Selection, Category: Health + Care; Hugo-Häring-Award 2011; Federation of German Architects /BDA - Baden-Württemberg

建筑师：韦伯纳阿拉扎德建筑与设计
项目地点：德国，尼尔廷根
总面积：24600平方米
建筑面积：3860平方米
完成时间：2010年
摄影师：©格尔德·尤塔图片设计公司
所获奖项：2012全球室内建筑设计奖；全球最佳室内和建筑设计奖（类别：健康+护理）；2011雨果-哈林奖；德国建筑师联盟/BDA巴登-符腾堡州

The special challenge of the task for the architects was the sensible use of the landscape with expansive views to and from the new building. They were looking for an architecture that fits in the landscape in consideration of materiality and structure, which is not hidden and rather represent itself powerfully and confidently as a place of healing to the outside world.

The great mass of the new building, approximately 25,000 cbm is so divided that the building not only harmoniously integrated into the landscape, but also meets the functional and in particular the atmospheric demands for a specialised hospital in an optimal way.

The new building has differentiated respond to the existing slope. The "stepped" elevated base level which responds to the slope houses all functional rooms and therapy rooms as well as the common areas. A continuous and well oriented magistrale that is parallel to Siebenkreuzerweg with an open view to the monastery Siessen connects all levels and functions of the building.

The elevated base level as the main artery of the new building can be experienced through the magistrale with good

1. Courtyard with landscaping
2-4. Exterior view from the yard
5. Façade detail
1. 进行景观美化的庭院
2~4. 从庭院看建筑外观
5. 建筑外立面细部

orientation as a contiguous space although it is split over three levels. The magistrale concentrates on the one hand the patient flow which is made possible by open views to the outside an easy orientation within the building; on the other, it forms a lively and informally meeting space for patients and therapists.

The architects have split the living quarters of the patients into three clear cubic units which are placed on the elevated base floor. This forms a clear readability of the functional and residential areas and it also creates a desired architectural proportion of the entire building. The residential areas are connected directly to the magistrale. French windows in every room allow one to have an open view to the beautiful landscape of Upper Swabia. The residential towers on the base level are twisted and shifted so that they do not cover each other and guarantee the residents of all living areas an open view to the landscape. A natural cladding made of local larch wood that will have a silvery gray colour in the course of time helps to ensure that the large building is taken as part of the landscape.

6. Distant view of exterior
6. 外观远景

East Elevation
东立面图

West Elevation
西立面图

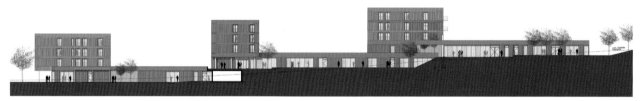

North Elevation
北立面图

South Elevation
南立面图

Floor Plans (Upper Three)
平面图（上方三图）

Section
剖面图

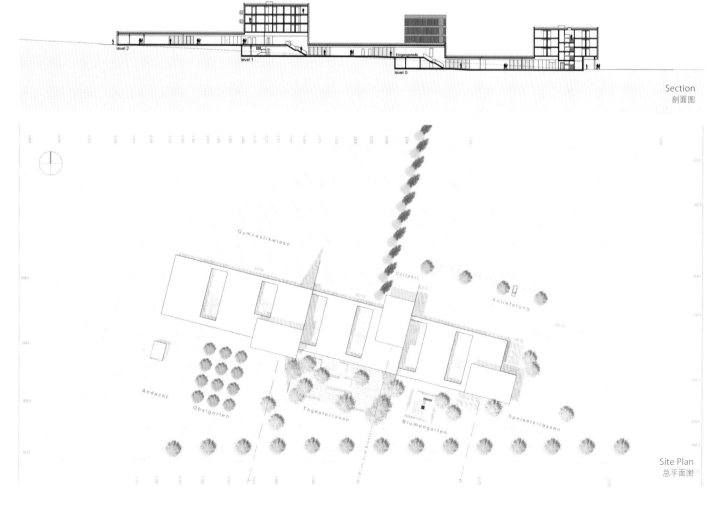

Site Plan
总平面图

7. Perspective view of corridor
8. Staircase
7. 走廊透视图
8. 楼梯

这个项目中最大的挑战是如何合理利用当地景观，优化视野。

设计团队力求设计出一座在材料和结构方面都与当地景观紧密融合的建筑，向外界呈现出积极有力的形象。

新楼25000立方米的设计方案不仅成功地使建筑与环境和谐相融，还满足了专项医院对建筑功能，特别是环境的严格要求。

建筑设计依照地势的变化而变化，"阶梯式"的底层设计满足功能室、治疗室和公共区域的不同需求。底层走廊通向各个楼层的不同功能区，并俯瞰Siessen修道院。

底层是新建筑的主要通道，底层走廊尽管分布在三个楼层，但朝向良好，空间连续。一方面走廊视野开阔，照顾病患在走动过程中对视野的要求；另一方面为患者和医生提供一个轻松惬意的见面场所。

病人的生活区被划分成三个整齐的正方体，分列底层之上。这样的设计不仅使功能区和居住区清楚地分开，还实现了理想的建筑比例。居住区与走廊直接相连，房间里的落地窗使住户可以欣赏到上施瓦本地区的美丽风景。居住区楼体有一定程度的扭曲移位，保证每个房间都拥有开阔的视野。使用当地木材制作的环保覆层将会随着时间的流逝呈现银灰色的格调，使建筑更好地融入当地的环境。

9. Entrance lobby
10. Staircase
9. 入口大厅
10. 楼梯

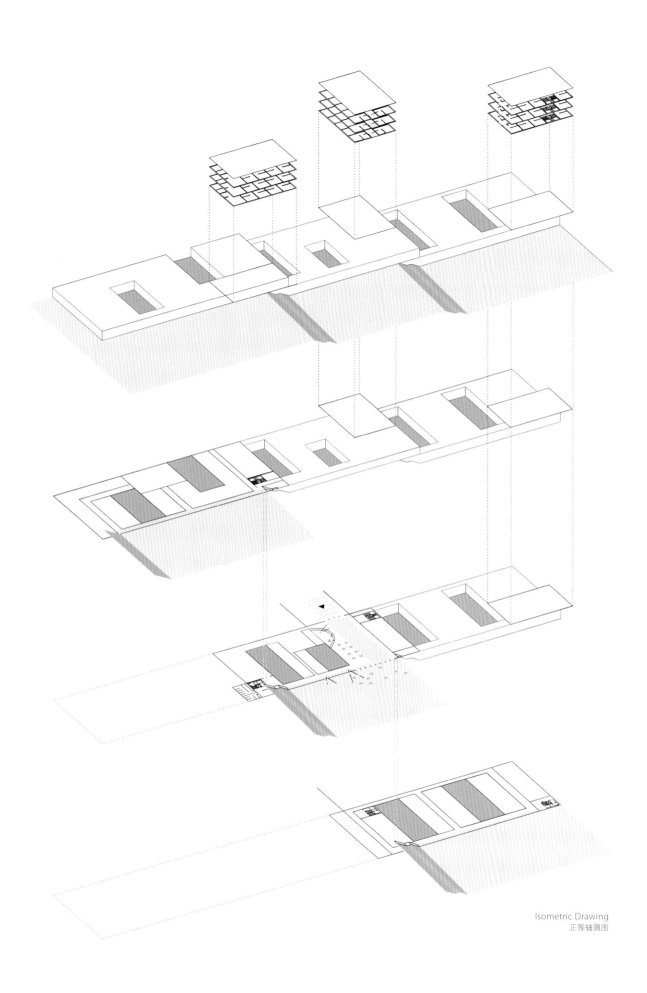

Isometric Drawing
正等轴测图

Boswijk Home
布什维基老年之家

Architects:
Dutch Health Architects
Location:
Vught, Netherlands
Building Area:
8,024m²
Project year:
2010
Photographs:
© Courtesy of EGM architecten
建筑师：荷兰健康建筑设计事务所
项目地点：荷兰，福格特
建筑面积：8024平方米
完成时间：2010年
摄影师：©EGM建筑公司

Boswijk Home offers accommodation and healthcare to elderly people suffering from dementia. In close collaboration with the client, EGM architecten developed a plan that answered to the sensitive nature and non-traditional characteristics of the residents. Patients with dementia respond to external stimuli. Therefore, the design of all the rooms and routes was carefully directed. The entire programme is on the ground floor. This gives a clear view of the tranquil green surroundings and invites the residents from their rooms out into the garden. Thus promoting the natural stimuli that are so important to these patients.

Patients can retreat into their own rooms whenever they wish, and live with other people together as one big family in a home with a private entrance and its own (living) environment. Twelve "stone" houses are spread out in a natural way over the site, connected to one another by small streets and plazas. The common facilities are housed in "stores" that form the walls in the street and plaza regions. The result is a complex in which the presence of the care organisation is almost invisible, and space is provided for the individual's quality of life. Home Boswijk was nominated for the 2010 Construction Plume (BouwPluim).

1. Courtyard viewed from interior
2. Side view of exterior with surrounding landscapes
3. The building is surrounded by green landscape and there are seats for the residents of the home
4. Approach to the building is surrounded by lawn

1. 从室内看庭院
2. 外观侧景及周围景观
3. 建筑被绿色景观围绕,其中有为居住者准备的座椅
4. 通往建筑的小路被草坪围住

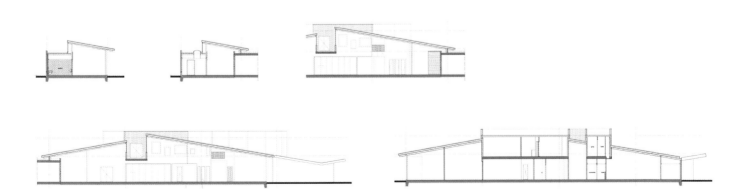

Sections and Elevations
剖面图和立面图

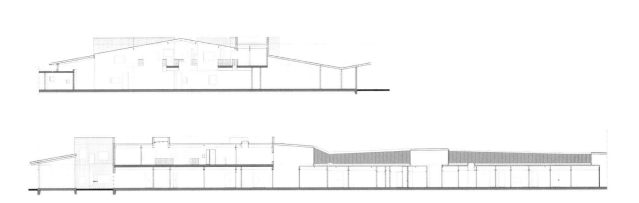

Sections
剖面图

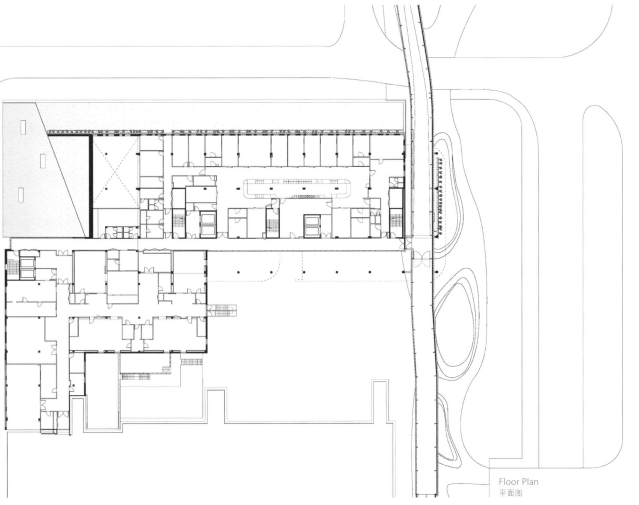

Floor Plan
平面图

5. Entrance viewed from the courtyard
6. View of corridor
5. 从庭院看向入口
6. 走廊内景

布什维基老年之家为患有老年痴呆症的老人提供住宿和医疗护理。EGM建筑公司从实际出发,针对老年之家敏感的非传统受众制定出了一个特殊的规划设计。老年痴呆症患者对外界环境较为敏感,因此老年之家所有的房间和路线都做了醒目的指示。单层的建筑设计方便欣赏室外宁静的自然环境,也吸引住户到室外花园走动,增加自然环境对病人的有益作用。

病人可以随时回到自己的房间,也可以与他人共享这个环境舒适的大家庭。12个"石头"房子随意地分散在各处,通过小树和广场彼此相连。街道和广场的围墙旁边是一些常用设施。老年之家在这样的环境中若隐若现,最大程度地保障住户的生活质量。布什维基老年之家获得了2010羽毛建筑大奖的提名。

7. Café/dining area

Facilities for Rehabilitation
康复设施

Clinica San Pablo Chacarilla Physical Therapy Centre
圣巴勃罗查卡利亚物理治疗中心医务总部

Architects: metropolisperu
Location: Lima, Peru
Building Area: 830m²
Project year: 2013
Photographs: © Courtesy of San Borja District in Lima

建筑师：秘鲁大都市设计公司
项目地点：秘鲁，利马
工程面积：830平方米
完成时间：2013年
摄影师：©由利马圣博尔哈区提供

The proposal contemplates the development of a Physical Therapy Centre and the clinic headquarters, all over an 830m² plot at San Borja District in Lima. The Centre has 5 floors and 5 parking basements. The building has 5 different vertical circulation systems: an elevator that connects all the levels, and stretcher elevator that goes from the 1st basement to the 3rd floor, 2 evacuation stairs that serve all the levels and 1 stair that connects the ground floor with the 1st.

The Physical Therapy Centre is compose in the following way:

Ground Floor
At the principal façade it has fit out four parking spots and the access ramp that leads you to the basements.

In front of the main entrance there is the admission area and the elevators and stairs that connect all the levels of the building. At your right hand there's the Centre Drugstore follow by the exit that leads you to the lateral entrance.

From here you can also enter the preceding part where the treatment pool, sauna, Turkish bath, hydro massage showers and other require services are found (also the patients and workers bathrooms). In here

1. Main entrance
2. Perspective view of lobby and reception
3. Lobby and reception
4. Physio gym and reception

1. 主入口
2. 大堂及接待处透视图
3. 大堂与接待处
4. 理疗健身室及接待处

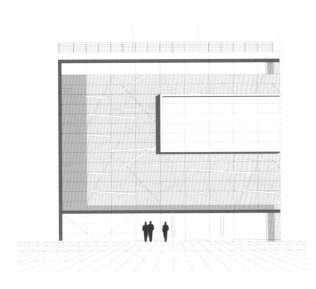
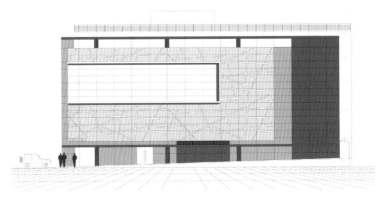

Elevations
立面图

there's a stair that connects only with the second level.
At your left hand, from the preceding part, there's also the evacuation stair and a corridor that leads you to an evacuation lateral exit.

First Floor
You access in here from the main evacuation stair, the elevator or the stair of the pool area. This level has two differentiate areas, one for doctor's offices and the other for a medical treatment room.

Second Floor
This level is exclusive for the Physical Therapy Centre's Gym. This area is destined for training machines, an aerobic exercise room and a spinning room. There is a central space that communicates the gym to the dressing rooms, bathrooms, showers and lockers. In here there is also a small food store for the public.

Third Floor
This level is exclusive for the Cardiovascular Recovery area. This area is destined for machines and equipment of cardiovascular rehabilitation, four doctor's offices and bathrooms. There is also a central waiting room that connects all the different areas.

Fourth Floor
In here the administrative area works. The manager's office, the directory, the support area and all the require services needed to attend the administrative personnel. Each of the offices has its own bathroom.

Parking Basements
Five parking levels have been set out. These basements are half levels connected by a ramp system. The access to these levels is through the elevators, the stretcher elevator and both of the stairs. The tank and the pump room are at the last basement.
The project counts with 75 parking spaces well distributed in the basements and 4 in the ground floor. 4 of this spots are reserve for handicapped people.
The ventilation system of all the bathrooms in the building will be solved by mechanic extraction system. Also the building will count with Detection System, Fire Alarms, Signposting, Emergency Lightning, Extinguishers, Cabinets and an Automatic Sprinkler System.

这个物理治疗中心医务总部的建设方案选址在利马圣博尔哈区一块830平方米的区域内。治疗中心的楼体结构包含地上5层和做停车场使用的地下5层。内部的垂直方向有5个流通空间：一部连接各层的电梯，一部可以到达地下一层至地上4层的担架电梯，两部连通所有楼层的疏散通道和一部连接1、2楼的楼梯。

物理治疗中心由以下结构组成：

一楼

主楼外部设有4个停车场和入口坡道，方便来访者到达地下车位。主入口前面是接待处和通往各层的电梯、楼梯。进门右手边是中心的药局，再往里便是连通外侧的出口。这里通往处理池、桑拿房、土耳其浴区、水力按摩房和其他相关设施（包括供病人和医护人员使用的卫生间）。这儿的楼梯只连通二楼。入口的左手边也有疏散通道，横向走廊通向出口。

二楼

从主疏散通道、电梯和处理池的楼梯都可以到达二楼，这一层分为两个主要区域：医生办公室和医疗处置室。

三楼

这层全部是物理治疗中心的健身房，设有训练

5. View of corridor
6. Staircase
7. Swimming pool
5. 走廊
6. 楼梯
7. 游泳池

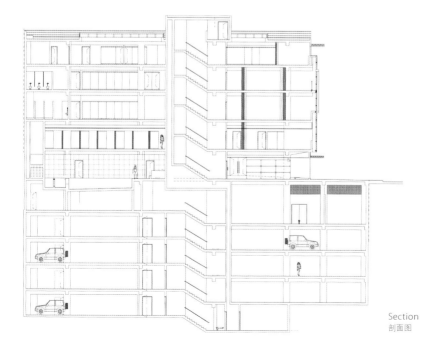

Section
剖面图

机械,有氧运动室和动感单车房。健身区域通过中央活动空间与更衣室、卫生间、淋浴间和储物柜相连。这里还有一个对外开放的小型食品商店。

四楼
中心西楼为心血管恢复区,安装了用于心脏康复训练的机械和设备,还设有4个医生办公室和厕所。各个区域由楼层中央的等候区相连。

五楼
五楼是行政管理区,包含经理办公室、检索目录、辅助设备和其他为行政人员提供的工作设施。每个办公室有独立卫生间。

地下停车场
治疗中心的地下设有5层停车设施,这些地下空间和半层结构由一个斜坡系统连接。通过普通电梯、担架电梯和两个楼梯都可以到达这里。水槽和泵室位于地下室的最底层。该治疗中心共能容纳75个车位,均匀地分布在地下停车场和地上一层,并设有残疾人专用车位。楼内卫生间的通风系统采用了机械抽取式设计,此外大楼还安装了计数检测系统、火灾报警器、路标指示牌、紧急照明装置、灭火器、壁橱和自动喷水灭火系统。

8. Multifunction room – meeting and education
9. Detail of bathroom
8. 多功能室——会议与教学
9. 盥洗室细部

Floor Plan – Level 1:
1. Main entrance
2. Admission and box
3. Administrative
4. Bathroom for the employed
5. Shower
6. Hiker
7. Whirlpool – whole body
8. Shower gingham
9. Pool
10. Men's dressing
11. Women's dressing
12. Dry sauna

一层平面图：
1. 主入口
2. 验票及售票
3. 管理处
4. 雇员盥洗室
5. 淋浴间
6. 徒步训练
7. 全身涡流浴池
8. 淋浴隔间
9. 泳池
10. 男士更衣间
11. 女士更衣间
12. 干式桑拿

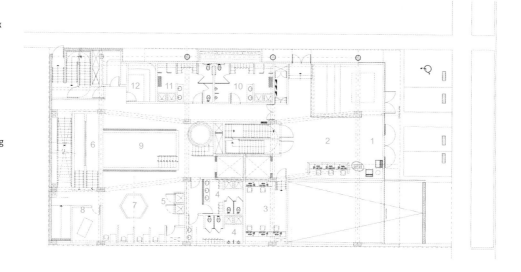

Floor Plan – Level 2:
1. Reception
2. Hairdressing
3. Manicure and pedicure
4. Treatment room
5. Staff room
6. Assistants counter
7. Consultant room

二层平面图：
1. 接待处
2. 美发
3. 修指甲与修脚
4. 治疗室
5. 员工室
6. 助理台
7. 咨询室

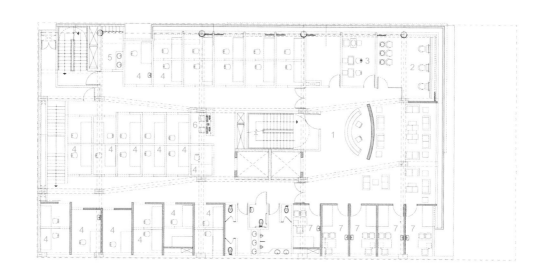

Floor Plan – Level 4:
1. Reception
2. Cardiovascular rehabilitation
3. Women's dressing room
4. Men's dressing room
5. Waiting space
6. Cardiovascular rehabilitation
7. Toilet
8. Consultant room

四层平面图：
1. 接待处
2. 心脏血管疾病康复
3. 女士更衣间
4. 男士更衣间
5. 等候区
6. 心脏血管疾病康复
7. 卫生间
8. 咨询室

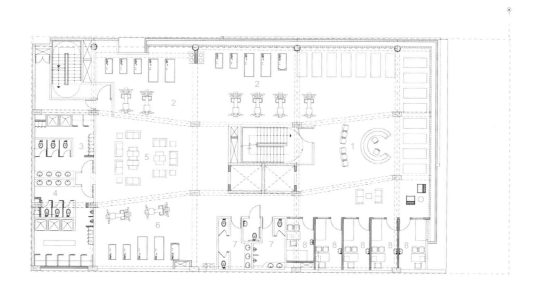

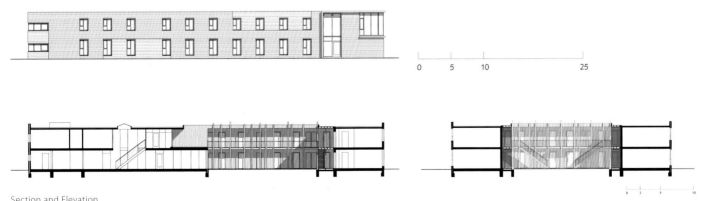

Section and Elevation
剖面图和立面图

Clinical Home Oegstgeest
乌赫斯特海斯特医疗中心

Architects:
Dutch Health Architects
(de Jong Gortemaker Algra)
Location:
Oegstgeest, the Netherlands
Construction Area: 5,000m²
Site Area: 1,500m²
Project year: 2012
Photographs: © Iemke Ruige
建筑师：荷兰医疗建筑设计师事务所
项目地点：荷兰，乌赫斯特海斯特
占地面积：5000平方米
建筑面积：1500平方米
完成时间：2012年
摄影师：©埃穆克·瑞格

The building contains 48 one-room apartments for people with a mental disorder. It is part of a psychiatric centre with a number of buildings set in an old country estate. The residents can handle a certain degree of autonomy, but are in need of quietness and protection at the same time.

The design meets these requirements by providing a protected environment for living under supervision, while maintaining freedom and privacy. Behind the brick facades of the two-storey urban block lies a concealed inner world with a different character.

The heart of the building is a courtyard, around which the apartments are grouped. For the residents, psychiatric patients who require structure and protection, we created a place with a warm and peaceful atmosphere.

The idea of the apartments around the courtyard refers to the classic Roman Villa with a central Impluvium. A cloister in two levels surrounds the space and contains the entrances to the apartments. The residents have their own private world with a tranquil atmosphere helped by the use of uniform and natural materials.

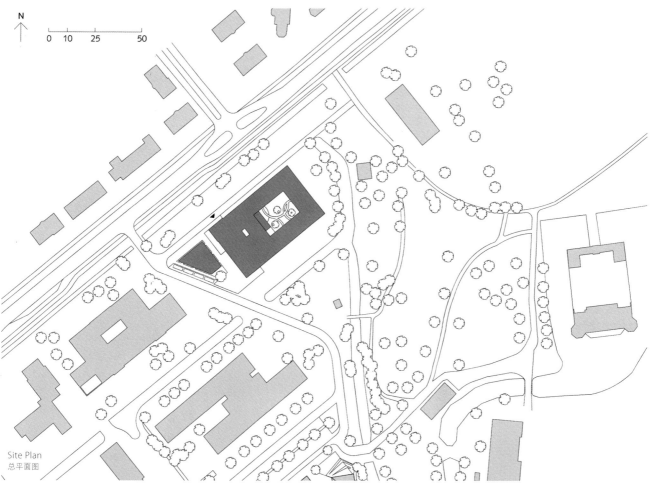

Site Plan
总平面图

1-2. Building and the surroundings
3-4. Courtyard with landscaping
5. Interior detail
1、2. 建筑与其周围环境
3、4. 庭院及景观美化
5. 室内细部

Urban planning conditions demanded the exterior to be a red brick volume. To make the bridge from an urban to a human scale the brickwork was designed with fine detail. Three different sizes of bricks were used to give the building a linear horizontal expression. In contrast with the exterior the courtyard is made entirely out of wood: sustainable hardwood planks for the facades and wooden beams and columns for the gallery structure.
The round columns are used in different lengths by analogy of trees. All the timber is untreated and will therefore form a natural gray patina by aging.

　　这栋建筑为精神障碍患者提供48间一居室公寓,是坐落在古老乡村庄园的一组精神病治疗中心的一部分。在这里居住的人能在一定程度上实现自我管理,但需要安静和稳定的环境作为保障。建筑设计在提供监管的同时,也保证住户的自由和隐私权力。在这个两层高的砖墙背后人们会发现一个截然不同的世界。

　　庭院是这个建筑的心脏,所有公寓都围绕院子排列。考虑到精神病患者对居住条件和安全感的需要,建筑设计强调温暖平和的氛围。公寓围绕院子这一结构安排模仿罗马别墅的设计理念,楼内回廊结构将建筑包围,并将公寓入口包含在内。使用统一的天然建筑材料确保住户拥有宁静的私人世界。

　　城市规划相关条款规定该建筑外部为红色砖结构。为了使城市环境与人居环境自然过渡,红砖外墙设计注重细节的雕琢。工程共使用了3种不同大小的红砖,增添建筑水平方向上的线性表现力。庭院结构全部使用木质建筑材料,与外墙形成鲜明对比。外层和横梁使用环保硬木木板,连廊结构使用木柱。不同高度的圆形木柱象征树林。木质材料均未经处理,随着时间推进将逐渐形成自然的灰绿色外表。

6. Dining area
7. Food counter
8. Dining room viewed from reception
6. 就餐区
7. 付餐台
8. 从接待处看餐厅

9. Kitchenette
10. Staircase
11. Meeting room
9. 小厨房
10. 楼梯
11. 会议室

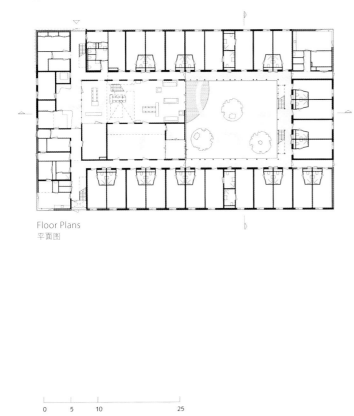

Floor Plans
平面图

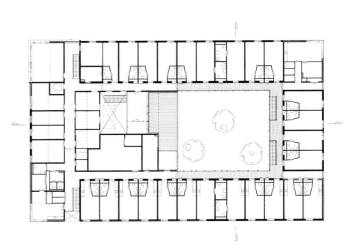

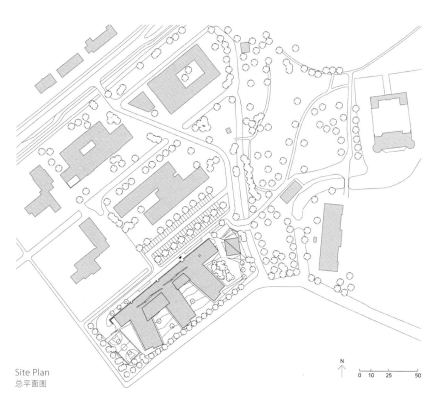

Site Plan
总平面图

Mental Health Care: "High Care"
"高度关怀"心理健康中心

Architects:
Dutch Health Architects
(De Jong Gortemaker Algra)
Location:
Oegstgeest, the Netherlands
Site Area: 8,200m²
Construction Area: 3,012m²
Project year: 2012
Photographs:
© Iemke Ruige, Tycho Saariste
建筑师：荷兰医疗建筑设计师事务所
项目地点：荷兰，乌赫斯特海斯特
占地面积：8200平方米
建筑面积：3012平方米
完成时间：2012年
摄影师：©埃穆克·瑞格、
第谷·萨里斯特

The Mental Clinic High Care is a closed facility accommodating people with severe psychiatric difficulties. The design contributes to the improvement of their situation by concentrating on the experience and relationship between inside and outside. This is made possible by introducing daylight into the building core, unorthodox corridors, and interaction with the landscape through views and use of materials. These interventions add up to a greater well-being for the clients thus resulting in a dramatic reduction of patient aggressiveness and less need for separation.

The design has been made in close collaboration with the staff and psychiatrists. Their objective was to give patients a pleasant and protected environment, safe for themselves and for others. Schizophrenia with aggressiveness or suicidality are common among the 36 patients of High Care. The architects' task was to provide both patients and personnel with a building, safe for both parties, while remaining a nice place to live.

The first step in reaching the goals was carefully placing the project on the site, an old country estate, resulting in a layout where the patient's quarters are placed

1. Rendering of the whole building
2. Entrance viewed from roadside
3. Patient room
1. 建筑整体效果图
2. 从路边看建筑入口
3. 病房

like fingers in the landscape. At the same time all the staff rooms are concentrated towards the street, protecting the patients' quarters from the outside world. This own protected world is further enhanced by transforming circulation from depressing corridors into bright and pleasant meandering routes, where clerestories allow for daylight into the building core.

The design provides direct views towards the gardens and landscape from anywhere in the building. The meandering shapes have become an integral theme, recurring in the walls, floor patterns and facades, creating a peaceful and characteristic habitat. The design succeeds in meeting staff requirements to be able to monitor patients, without sacrificing their privacy and freedom of movement within the boundaries of the building.

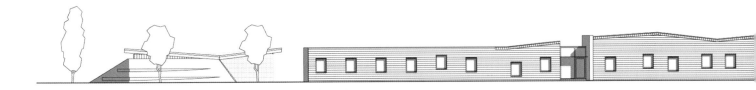

3

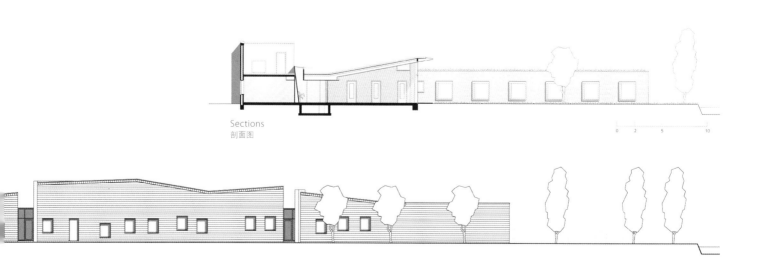

Sections
剖面图

4. Exterior viewed from courtyard
5. Lobby with lounge/meeting area
4. 从庭院看建筑外观
5. 大厅配有休息/会谈区

"高度关怀"心理健康中心是一间收容严重精神病疾病患者的全封闭机构。建筑设计强调内部与外部的关系与体验，有利于病人改善病情。通过将自然采光引入建筑中心空间，不规则走廊的设计以及视线视角和所用材料与周边环境进行互动。这些干预措施有益于患者的身心健康，对降低病人攻击性，减少隔离必要性都有不错的效果。

建筑的设计过程积极地争取到了工作人员和医生的参与配合，旨在为病患提供一个舒适而有安全感的环境，这对病人和工作人员都是有益的。心理健康中心的36名病人大多患有具备攻击性和自杀倾向的精神分裂症，这使得建造设计师面临的最大挑战就是在确保病人和医护人员双方安全的前提下建造一个舒适愉快的生活空间。

实现这个目标的第一步就是选址。设计师在经过了仔细的考察后决定将病房按发散的形状排列。医护人员的房间则朝向街道，在病房与外界环境之间形成一道屏障。放弃传统走廊，采用明亮开放的蜿蜒式走廊设计进一步加强了对病人空间的保护。这里的天窗极大地改善了建筑的采光条件。

从建筑的各个方位都可以欣赏到花园和周围的景观。蜿蜒的形状是整个园区的主题，在墙壁、地板花纹、外墙上都有所体现，营造出平缓和谐的环境特点。这个设计方案的成功之处在于满足了医护人员在不影响病人隐私和行动自由的前提下照看病人的要求。

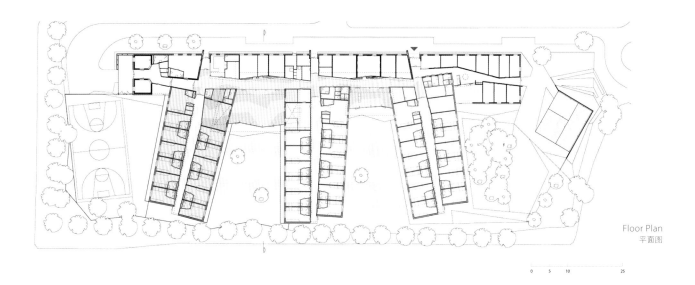

Floor Plan
平面图

6. Perspective view of corridor
7. Plant decoration in corridor
8. View of corridor
6. 走廊透视图
7. 走廊内的植被美化
8. 走廊

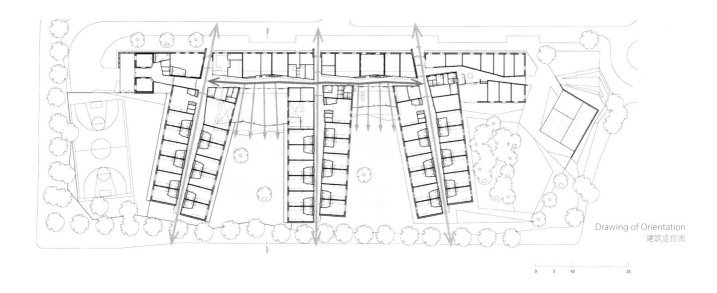

Drawing of Orientation
建筑定位图

The Volgerlanden

国家心理健康护理中心

Architects:
Dutch Health Architects
(de Jong Gortemaker Algra)
Location: Hendrik Ido Ambacht, the Netherlands
Site Area: 10,500m²
Construction Area: 5,050m²
Project year: 2011
Photographs: © Iemke Ruige

建筑师：荷兰医疗建筑设计师事务所
项目地点：
荷兰，亨德里克-伊多-阿姆巴赫特
占地面积：10500平方米
建筑面积：5050平方米
完成时间：2011年
摄影师：©埃穆克·瑞格

The Volgerlanden accommodates 60 clients, who are dependent on long-term psychiatric care. The layout of the building stimulates living in a social community, but also allows for independent living. Careful attention to scale, daylight, detail and material has resulted in an elegant, modest building which contains a surprise inside!

The building connects to the ecological zone that runs through a new residential area. The design for the landscape and selection of vegetation are a continuation of this ecozone.

The use of modest and natural looking materials was also inspired by the site.
Floor to ceiling windows divide the masonry facades in planes that correspond to the scale of the dwellings in the area, causing the building with its sloped sedum roof to naturally merge with its environment.

All clients have their own apartment with living room + kitchenette, bedroom, bathroom and a private terrace or balcony. The apartments are divided over the two levels of the building and have a large variation in orientation, meeting the specific

1

needs of the clients. The inner apartments are situated around secure patios, while the outer apartments either have a panoramic view of the neighbourhood, encouraging contact with the residential area, or face the tranquil canal.

The facility does not appear as a "psychiatric institution", but rather as a safe living environment that is spacious and light. The patios allow daylight into all corridors. The entrance hall of the building opens up into an atrium with a glazed roof, the heart of the building, which functions as an informal meeting area. Adjacent to the atrium are several collective facilities, such as a sport activity room, a restaurant and four lounge rooms with direct access to the patios. Each lounge room has its own theme and atmosphere, providing a variety of choice for the clients.

1. Overall view of exterior from waterside
2. Exterior winter view
3. Exterior detail viewed from patio
1. 临水建筑全景
2. 冬季外观
3. 从天井看建筑外观局部

Site Plan
总平面图

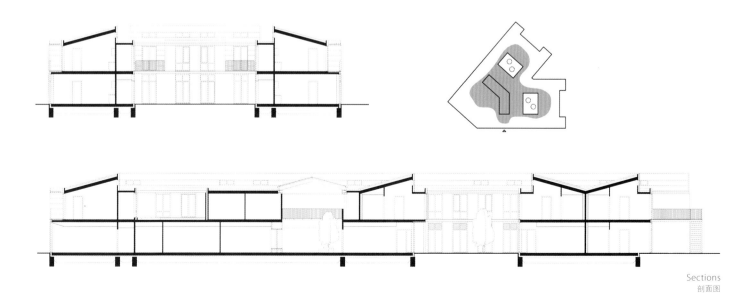

Sections
剖面图

155

4-7. View of lobby
4~7. 大堂

国家心理健康护理中心可以为60位精神病患者提供长期护理。建筑格局遵循社会社区的设计形式，但也提供独立的生活空间。建筑规模、采光通风、建筑细节和材料方面的设计将护理中心打造成一个外部精致美观，内部充满惊喜的建筑。

护理中心与贯穿一个住宅区的生态园区相连。这里的景观设计和植被选择都延续了生态园的风格和理念。这些低调、天然的建筑材料结合了当地的环境特点。落地窗将砖石外墙分隔成一个个与周边住宅相呼应的平面。倾斜屋顶也与周围环境和谐相融。

所有的患者公寓都有厨房与起居室为一体的客厅、卧室、卫生间和独立露台或阳台。公寓分布在建筑的两层，融合多种朝向设计，以满足病人的不同需求。朝内的公寓位于安全庭院周围，朝外的公寓要么能够欣赏到周围的景色，激发病人接触外部环境，要么面对宁静的运河。

建筑师对这座建筑的定位不是"精神病院"，而是一个宽敞明亮又安静和谐的居住环境。庭院的设计使走廊采光充足。入口大厅正对的中庭安有玻璃顶，是整个建筑的中心，作为非正式的会议室使用。中庭周围有许多公共设施，包括体育活动室、餐厅和4间直通院子的休息室。每个休息室都设计了不同的主题和氛围，为病人提供多种多样的选择。

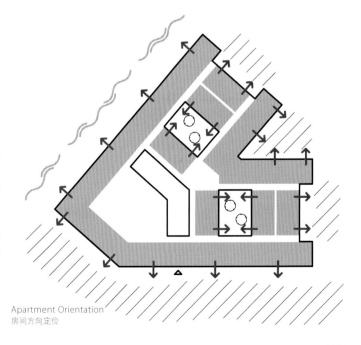

Apartment Orientation
房间方向定位

8-10. View of passage
8~10. 通道

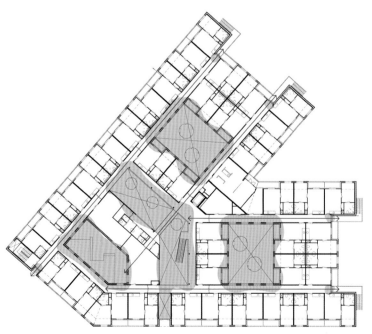

Sequence of Light – First Floor
采光序列——二层

11-12. Dining room
11、12. 餐厅

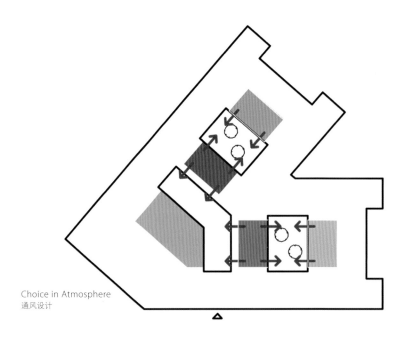

Choice in Atmosphere
通风设计

Ground Floor Plan
首层平面图

First Floor Plan
二层平面图

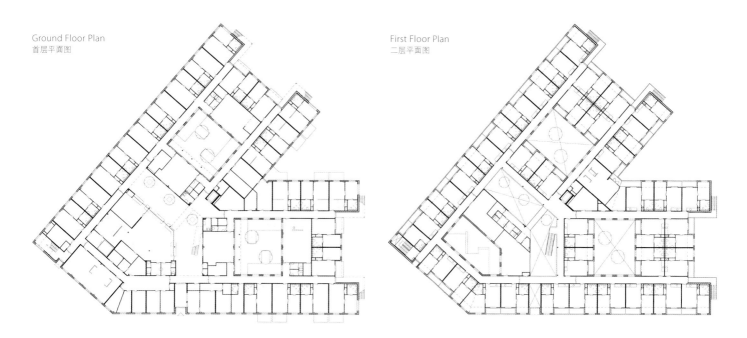

Extension of the Rehab Healthcare Facility
康复护理中心扩建工程

Architects:
d.A.p architecture & design
Location: Laterza, Italy
Total Surface:
15,779.41m²
Project year: 2010
Photographs:
© Courtesy of
d.A.p architecture & design
建筑师：d.A.p建筑设计事务所
项目地点：意大利，拉泰尔扎
总面积：15779.41 平方米
完成时间：2010年
摄影师：©d.A.p建筑设计事务所

Functional Rehabilitation Facility
The area concerned with the extension of the Rehab Healthcare Facility is south-west of Laterza, a city near Taranto. The region is characterised by the typical wonderful landscape of the western area of the "Murgia Tarantina". This area represents an exceptional match of natural and historical elements, with its geological phenomenon of the "gravina", a canyon that is up to 400m wide and 200m deep. The core principles of the project design are the integration into the region that is characterised by a high environmental and conservationvalues, and close attention to the living spaces that must assure all comforts for the patients. The horizontal development of the layout plan with three separate buildings and fluent shape that imitates the existing environment, stands for the intention to propose a homelike image with a low environmental impact. The paved external areas connect the central building units and produce a "continuum" between indoor and outdoor space emphasised by wide, glazed surfaces. The result is an architecture where the natural light and the environment are the main quantitative and qualitative elements in the indoor spaces. From 2006 up to 2010 the development

1. Entrance viewed from patio
2. Exterior viewed from courtyard
3. Exterior viewed from the garden
1. 从院子看向入口
2. 从庭院看建筑外观
3. 从花园看建筑外观

planning of the whole area concerned the parcelling out divided in 4 lots:
Lot I completely realised: 170 beds for functional rehabilitation;
Lot II in progress: 100 beds for functional rehabilitation of people with physical, mental and sensory disabilities and 40 beds in Residential Care Facilities (R.S.A.);
Lot III in progress: Hall, conference hall, offices, 60 beds for functional rehabilitation, 20 beds in R.S.A., 18 beds for palliative treatments, 14 beds for Residential Facilities for people suffering from psychiatric problems;
Lot IV hotel for people with a physical disability, including a therapeutic swimming pool, a recreational area and a conference hall. According to the project, it is expected the construction of facilities for a total volume of 106,355.50m^3, a total surface of 15,779.41m^2, 10,736.97m^2 of parking area, 27,000.62m^2 of green.

The first lot is developed in three levels:
Multifunctional area on the ground floor with hospitalisation rooms and general services: lobby and reception; administrative offices and worship rooms. Hospitalisation area on the first floor with rooms and relative services: living-room, small kitchen, dining-room, toilets and aided bathrooms for disabled people and service areas. Service area on the second floor with therapies, gyms, dressing rooms, toilets and storage areas.

4. Courtyard viewed from balcony
5-7. View of interior
8. Reception
4. 从阳台看庭院
5~7. 室内内景
8. 接待处

康复中心的功能设施

距离塔兰托不远的拉泰尔扎西南地区是这项康复中心扩建工程的所在地。这里拥有"塔兰托–穆尔吉亚"西部地区的典型地貌以及称为"格拉维纳"的地质现象，即400米宽、200米深的大峡谷。自然与历史在这里实现了独特的融合。工程的基本原则是使建筑充分融入其具有高度环保价值的环境中去，同时保证所有病人都能拥有舒适的生活空间。

该地区的最大特点就是"塔兰托–穆尔吉亚"西部地区的典型地貌。项目的横向规划设计包含3座独立建筑和模仿当地地貌的流畅造型，旨在保持较低环境影响的前提下打造出熟悉亲切的生活环境。建筑外部的路面连接中央的建筑单元，通过宽阔玻璃墙面的使用在室内和室外形成"统一体"。自然采光和周边环境是影响室内的主要定量和定性元素。

2006年到2010年间，整个工程区域的发展规划分为4个部分：

地段1.完全实现：用于功能康复的170个床位。

地段2.进行中：用于生理、心理和感官残疾功能康复的100个床位和用于长期护理的40个床位。

地段3.进行中：大厅、会议厅、办公室，用于功能康复的60个床位，用于长期护理的20个床位，用于舒减治疗的18个床位，用于精神病患者长期护理的14个床位。

地段4.方便残疾人使用的酒店，包含治疗性泳池，休闲娱乐区和会议厅等设施。

竣工后的总建筑空间预计可以达到106,355.50立方米，总建筑面积15,779.41平方米，停车面积10,736.97平方米，绿地面积27,000.62平方米。

对地段1的开发是在3个层面上进行的：

一楼设有多功能活动区、住院处和一般服务区；一般服务区包括大堂和接待处，行政办公处和宗教礼拜室。住院区位于二楼，设有房间和配套设施，包括客厅、小型厨房、餐厅、卫生间，还有残疾人浴室和服务区。三楼的服务区提供治疗室、健身房、更衣室、卫生间和储藏室。

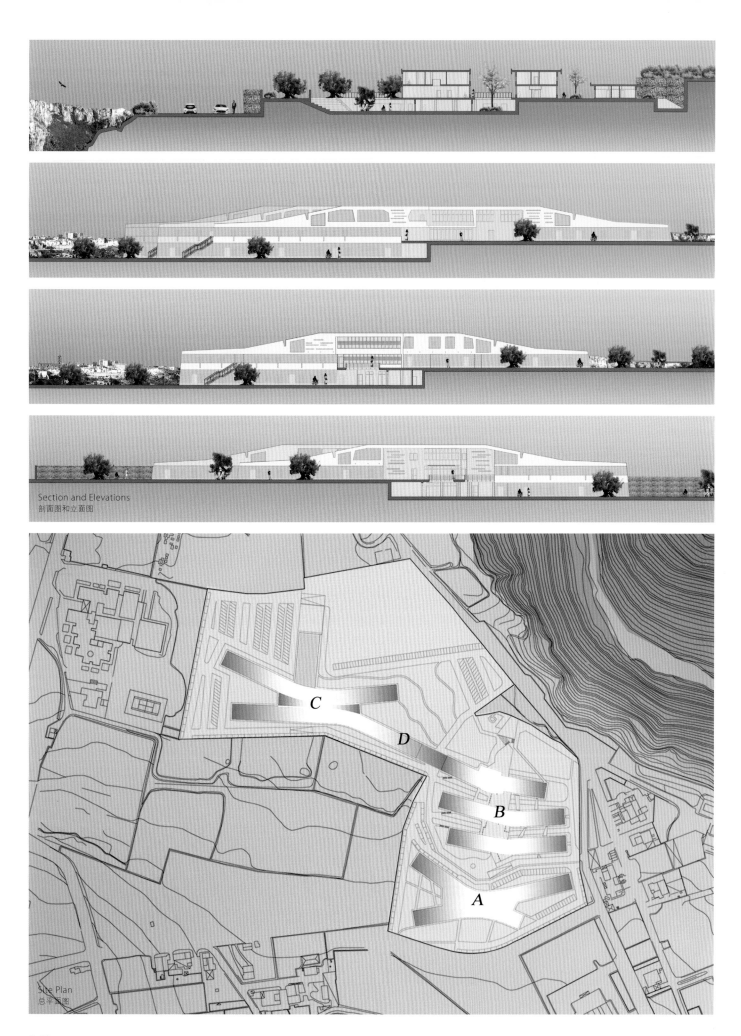

Section and Elevations
剖面图和立面图

Site Plan
总平面图

9-10. Rehabilitation gym
11. Dining area
9、10.康复治疗健身室
11.餐饮区

Cross Section:
1. Resting area
2. Main Pool Area
3. Terrace
4. Technical Area

横截面图：
1. 休息区
2. 主泳池区
3. 露台
4. 技术服务区

Longitudinal Section:
1. Sauna Terrace
2. Main Pool Area
3. Pool Terrace
4. Technical Area
5. Night Club
6. Massage Area

纵剖面图：
1. 桑拿露台
2. 主泳池区
3. 泳池露台
4. 技术服务区
5. 夜总会
6. 按摩区

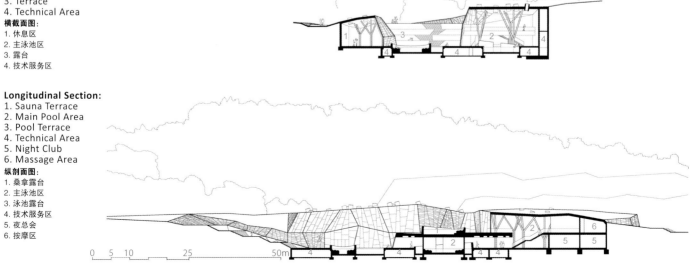

Wellness Centre Orhidelia
奥迪利亚疗养中心

Architects: Enota
Location: Podčetrtek, Slovenia
Site Area: 9,880m²
Construction Area: 5,030m²
Photographs: © Miran Kambič
Awards: Golden Pencil 2009; Contractworld Award 2010, shortlisted; World Architecture Festival Award 2009, shortlisted
建筑师：Enota建筑事务所
项目地点：斯洛文尼亚，博德森特克
占地面积：9880平方米
建筑面积：5030平方米
摄影师：©米兰·坎比克
所获奖项：2009金铅笔奖；2010 Contractworld室内设计奖，入围设计；2009世界建筑节，入围设计

Wellnes Orhidelia is situated in Terme Olimia thermal complex in Podčetrtek, Slovenia. Founded in late seventies, when they started to exploit thermal water, the complex was originally designed as health center for medical cure after injuries and the visit was "sponsored" by state health insurance company. With the fall of socialistic system, situation changed radically. The owners were forced to adapt to new conditions, trying to take advantage of beautifully surroundings and invite the customers with more wellness-oriented program.
Lately the new management is trying to lift the bar even higher and Wellness Orhidelia is third in a row of projects that enota was asked to do in last couple of years. It follows the extension of original pool complex Termalija and new Hotel Sotelia which was designed to attract also more demanding guests.
All three project were designed with an ambition to intervene as little as possible into the existing scenery trying to connect with beautiful nature of the area.
Main goal while designing the building was to diminish as much as possible its presence in the surroundings. Since the demanded program of wellness center is very extensive and in parts it demands overcoming great spans and big heights of inner spaces, putting

3D Scheme and Visualisation
三维图解和视觉示意图

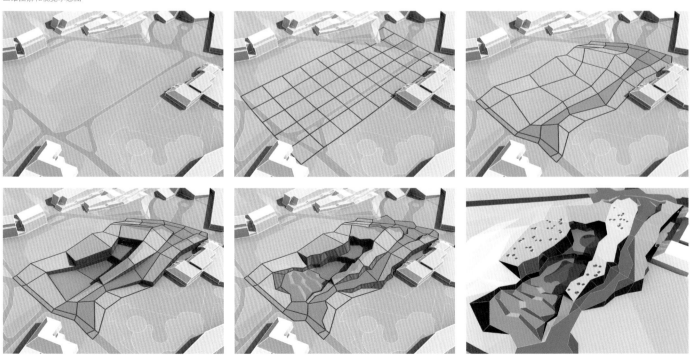

1. View over central atrium
2. Central atrium with pools and sauna terraces
3. Building is designed like a landscape arrangement
4. Main strolling path leads over the roof

1. 从中庭看到的外景
2. 中庭配有泳池和桑拿露台
3. 建筑的设计似景观布局
4. 主要的漫步通道越过屋顶

up classically conceived building on central green plot would fill up last remaining open area in thermal complex and largely degraded its spatial quality.

New wellness centre is consecutively designed rather like a landscape arrangement than a building. Folded elevations appear like supporting walls dividing different levels of landscape surfaces. Central walking path is now stretched over the roof and enables visitor completely new, different experience of the site. On both ends, where strolling path connects with passing inner roads, it forms two smaller public squares to control the speed of vehicles and ultimately gives advantage to pedestrians over the traffic.

Rather than searching for its own expression and claiming its space new object connects existing single buildings and other spatial elements in the whole.

The Wellness Orhidelia in Podčetrtek, designed by the Enota architectural office and Bruto landscape-architects, is the most obvious example of succesful merging of different disciplines. This project is architecture, landscape and intetior design at the same time, articulated in an outspoken geometrical vocabulary.

Although it is not known how the architect and landscape architect actually collaborated, it considers it anyhow of importance that this is an interdisciplinary project. Compared to the glitzy superficiality of many spas and resorts – in Slovenia and elsewhere – this complex reveals a setious concern for architecture.

5

奥迪利亚疗养中心位于斯洛文尼亚的博德森特克温泉胜地，始建于20世纪70年代末。当时人们刚刚开始利用温泉水，在当地建立了理疗中心，方便人们用温泉疗伤，也为国家医疗保险公司的赞助对象提供服务。随着社会主义制度的衰落，这里的情况也急转直下。温泉所有人不得不适应新情况，利用当地优美的自然环境和更加有益健康的项目吸引客人。

近来，当地的新管理层尝试进一步提高标准，奥迪利亚疗养中心成为Enota建筑设计事务所在几年间获邀开展的系列工程中的第三个项目。前面的两个工程分别是特玛利亚温泉疗养区的扩建项目和满足挑剔顾客的新索特拉酒店。

这三个工程的设计都尽量保证对原有环境的影响降到最低，并与大自然融为一体。

建筑设计中的主要目标是尽量减少对周围环境的影响。由于疗养中心涉及范围较广，局部结构需要挑战内部空间的跨度和高度极限，而包容挤占中央绿化带的传统建筑会将占用最后的空地，极大地降低空间质量。

新疗养中心的设计是连续进行的，更类似于景观规划而非建筑设计。多层次的立面设计像支撑墙一样将景观面的不同层次分隔。中央步行街一直扩展到屋顶之上，为来客提供前所未有的视角和体验。步行街的两端与公路相连，形成两个小型公共广场，对车辆的速度进行调节和控制，为行人提供安全与方便。

这个工程的特点在于选择与原有的独体建筑以及其他空间元素结合为一个整体，而不是过于强调独立的空间和个性。

博德森特克的奥迪利亚疗养中心由Enota建筑事务所的和布鲁托景观建筑公司设计，是不同领域合作的成功案例。疗养中心是一个兼具建筑设计、景观设计和室内设计的项目，采用了直白明快的几何语言。

尽管建筑设计师和景观设计师的合作过程不为人知，但其成果显然是十分丰硕的。与斯洛文尼亚其他地方的许多浮夸的温泉度假村相比，奥迪利亚疗养中心展示了深刻的建筑思考。

5. Inner pool and nature like supporting structure
6. Inside connecting with exterior at nightfall
7-8. Relaxation area
5. 室内泳池和仿自然支撑结构
6. 黄昏，建筑内外连在一起
7、8. 放松、娱乐区

Ground Floor Plan:
1. Resting room
2. Main pool area
3. Snack bar
4. Night club
5. Toilettes, shower
6. Toilettes
7. Technical area
8. Pool terrace

首层平面图:
1. 休息室
2. 主泳池区
3. 快餐吧
4. 夜总会
5. 化妆室,淋浴
6. 化妆室
7. 技术服务区
8. 泳池露台

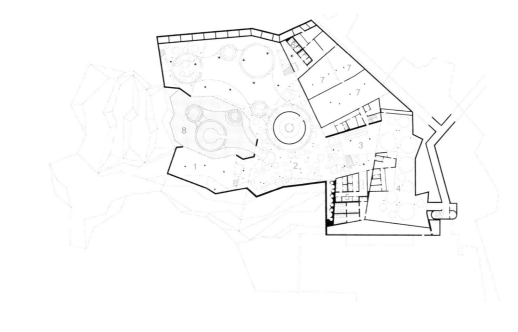

First Floor Plan:
1. Showers, toilettes
2. Resting area
3. Saunas
4. Steam bath
5. Waterbeds area
6. Massage area
7. Pool terrace
8. Sauna terrace

二层平面图:
1. 淋浴、化妆室
2. 休息区
3. 桑拿
4. 蒸汽浴
5. 水床
6. 按摩区
7. 泳池露台
8. 桑拿露台

Roof Plan:
1. Pool terrace
2. Sauna terrace
3. Main strolling path over the building

屋顶平面图:
1. 泳池露台
2. 桑拿露台
3. 主要漫步通道越过屋顶

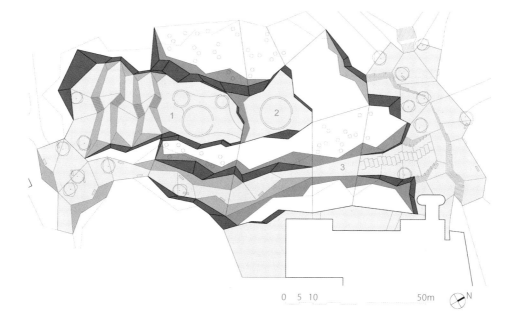

Tokyo Bay Rehabilitation Hospital
东京湾康复医院

Architects:
RTKL
Location:
Yatsu, Tokyo, Japan
Construction Area:
8,140m²
Photographs:
© RTKL.com/David Whitcomb
建筑师：RTKL建筑事务所
项目地点：日本，东京，谷津
建筑面积：8140平方米
摄影师：©RTKL.com/大卫·惠特科姆

Healthcare Japan commissioned RTKL to design an 8,140m² rehabilitation hospital for the campus of Yatsu-Hoken Hospital in Tokyo. Completed in 2007, the new facility houses a full rehabilitation gym, activity day living facilities, imaging facilities, clinics, support rooms, a hair salon, and a Japanese or Tatami room.

Maximising natural light, incorporating elements of nature, and creating flexible indoor spaces are primary characteristics of the building's design – all focused on enhancing the patient experience and providing for the needs of family and caregivers.

A cylindrical atrium space at the heart of the facility helped answer these project challenges. It introduces natural light into the center of the facility while allowing patient rooms to be arrayed to capture the light and ventilation.

The atrium space is complemented by an oval-shaped bath tower adjacent to the main entrance. The tower creates a wayfinding element since it can be clearly seen from the existing Yatsu Hoken Hospital.

To maximise space, as many areas as possible were designed to be used flexibly. For instance, the dining areas and day rooms on each floor can be opened to each other

1. Exterior viewed from street
2. Night view of exterior, main entrance
3. Day view of entrance
1. 从街道看建筑外观
2. 外观夜景，主入口
3. 入口

and used as additional rehab areas.
Another benefit of the open plan is the visual and auditory connection between patients and staff, ensuring that patients feel neither isolated nor institutionalised. While still in visual contact with caregivers, patients are encouraged to use the corridor with a view to the atrium for exercise. Not only is this beneficial to their treatment, it also provides a helpful distraction for those who tend to get disheartened by long hospital stays.

Healing gardens, located both at ground-level and roof-top-level, include plant containers set at an accessible height so patients can garden as part of their therapy. Other outdoor amenities include dining facilities that extend onto a terrace and balconies outside all patient rooms.

All in all, the aim in designing Tokyo Bay Rehabilitation Hospital was to optimise the healing environment for patients, families, and caregivers alike.

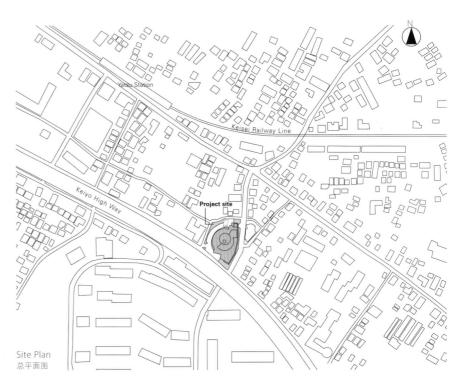

Site Plan
总平面图

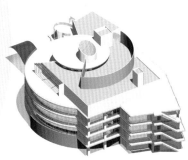

Axonometric (Upper Three)
轴测图（上方三图）

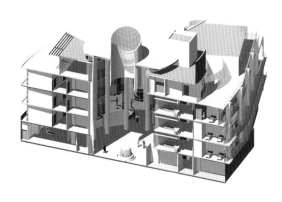

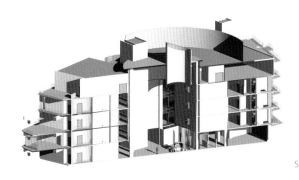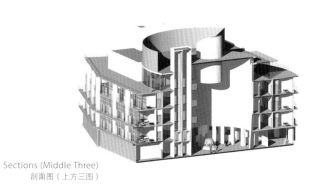

Sections (Middle Three)
剖面图（上方三图）

Elevations (Below Four)
立面图（下方四图）

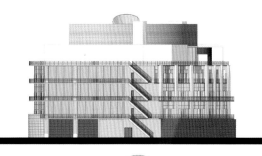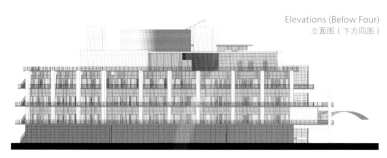
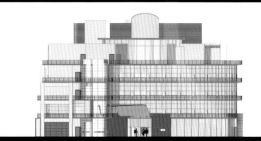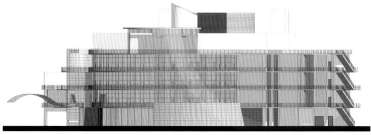

Site Plan – Landscape :
1. Service gate
2. Staff parking
3. Fence/screened
4. River rock paving
5. Architectural screened wall
6. Ornamental trees
7. Entry
8. Steel raised planters
9. Private garden
10. Bench
11. Special paving
12. Public walk
13. Steel raised planters
14. Fire tank

总平面图——景观美化：
1. 服务入口
2. 员工停车场
3. 栅栏/遮蔽体
4. 河卵石铺路
5. 建筑遮挡墙
6. 观赏树木
7. 入口
8. 钢结构体养种植物
9. 秘密花园
10. 长椅
11. 特殊块石铺路
12. 公用步行道
13. 钢结构体养种植物
14. 消防水罐

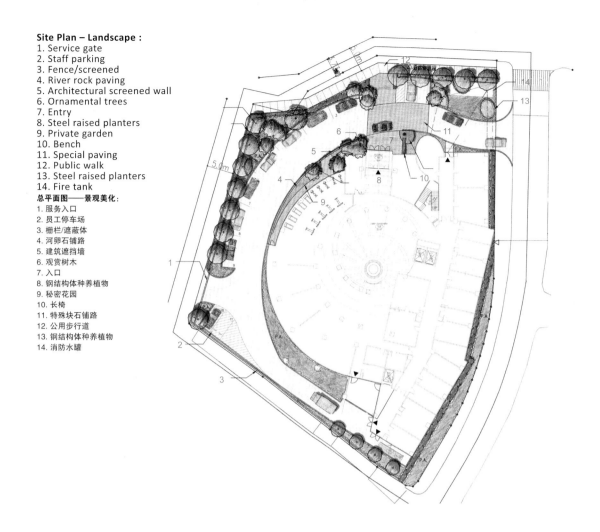

Roof Plan – Landscape:
1. Special paving/tie pavers
2. Therapy garden
3. Bench
4. Ornamental trees
5. Trellis structure
6. Special paving/tile pavers
7. Egress route path (emergency exit)
8. Therapy garden
9. Sedum mats/500x500

屋顶平面图——景观美化：
1. 特殊块石铺路/带状铺面
2. 治疗花园
3. 长椅
4. 观赏树木
5. 格架结构
6. 特殊块石铺路/带状铺面
7. 出口路径（紧急出口）
8. 治疗花园
9. 景天属植物区块/500x500

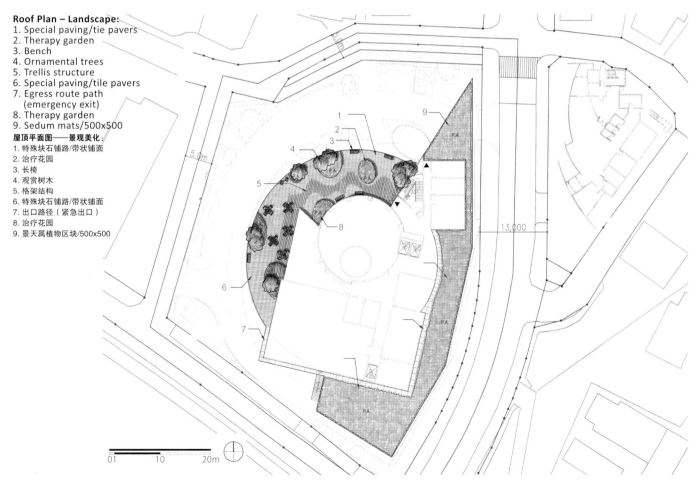

4. Lounge terrace
4. 休息露台

5. Atrium with light well
6. View of atrium
5. 中庭及采光孔
6. 中庭

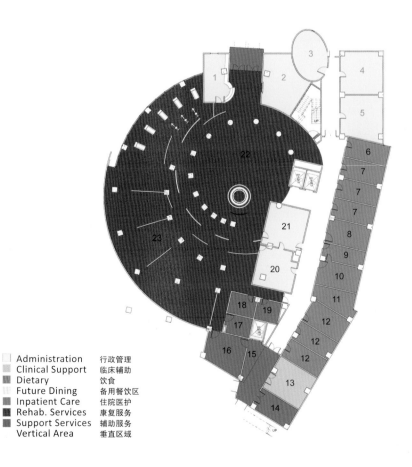

Ground Floor Plan:
1. Reception
2. Office
3. Security
4. Staff lockers
5. Nurses lockers
6. Treatment
7. Exam
8. X-ray
9. Control
10. CT
11. Bone dens.
12. Cardio/lab
13. Sterile service
14. Morgue
15. Service area
16. Dispensary
17. Store
18. Female WC
19. Male WC
20. Social service
21. Staff office
22. Lobby/lounge/kiosk
23. Rehabilitation service

首层平面图：
1. 接待处
2. 办公室
3. 安保
4. 员工更衣室
5. 护士更衣室
6. 治疗室
7. 检查室
8. X光照片室
9. 控制室
10. CT室
11. 骨密度室
12. 心脏/实验室
13. 消毒服务
14. 太平间
15. 服务区
16. 药房
17. 储藏室
18. 女士卫生间
19. 男士卫生间
20. 社会服务
21. 员工办公室
22. 大厅/休息/亭台
23. 康复服务

Administration 行政管理
Clinical Support 临床辅助
Dietary 饮食
Future Dining 备用餐饮区
Inpatient Care 住院医护
Rehab. Services 康复服务
Support Services 辅助服务
Vertical Area 垂直区域

7-8. View of corridor
7、8. 走廊

RTKL建筑事务所受日本医疗协会的委任在东京谷津保险医院进行总面积8,140平方米的康复医院设计。项目于2007年竣工，包含了完整的康复健身中心、短期活动设施、成像设施、诊所、储藏间、美发沙龙以及日式榻榻米房间。

康复中心的基本设计特点是最大程度加大采光，融合自然元素，创造灵活的室内空间。在改善病患就诊体验的同时为病人家属与陪护提供方便。

建筑中央的圆柱形中庭设计满足了以上所有需求。不仅可以将自然光引入建筑的中心，还优化了病房的采光通风条件。

靠近建筑主入口的椭圆形浴塔设计使中庭空间更加完整。在旁边的医院就可以看到这个塔形结构，方便行人辨别方向。

为了增大室内空间，提高利用率，许多空间都采用了灵活设计。举例来说，每个楼层的用餐区和日间活动室都可以相互连通，作为额外的康复区域使用。

开放设计的另一个好处是利于病人与医护人员进行视觉和听觉上的互动，确保病人既不会感到孤独，也不会感到单调无聊。在陪护人员的监护下，病人可以在走廊走动锻炼，同时欣赏中庭的景色。这不仅对病人的治疗有益，也利于陪护人员在长期的医院生活中获得短暂的放松。

治疗花园分布在建筑一层和顶层，设有高度适宜的花盆，方便病人参与园艺活动，加强治疗效果。另外，病人房间的露台和阳台上还设有餐饮设施。总而言之，东京湾康复医院的设计理念即优化治疗环境，方便病人、病患家属、医护人员以及一切相关人员。

First Floor Plan:
1. Bath
2. Speech therapy
3. Four-bed ward
4. Pantry
5. Day room/dining
6. Nurse station
7. Rehabilitation
8. Equipment
9. Treatment
10. Clean utility
11. Soiled utility
12. Laundry
13. Atrium
14. Private room

二层平面图：
1. 浴室
2. 语言障碍矫正
3. 四人间病房
4. 食品室
5. 休息室/餐饮
6. 护士站
7. 康复
8. 设备间
9. 治疗室
10. 清洁品杂物间
11. 污物间
12. 洗衣间
13. 中庭
14. 私人间

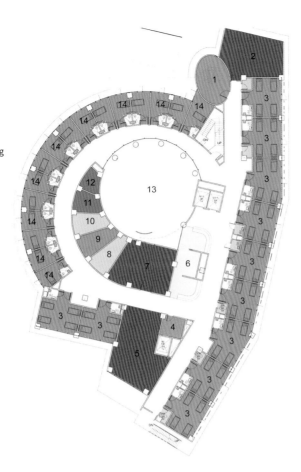

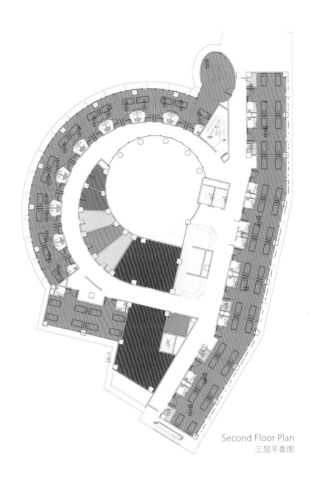

Second Floor Plan
三层平面图

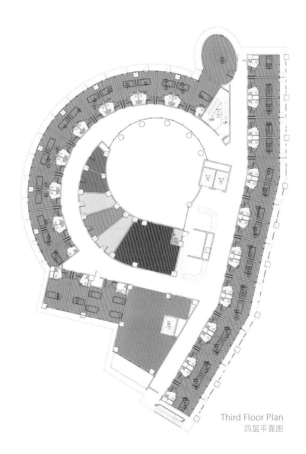

Third Floor Plan
四层平面图

Administration 行政管理
Clinical Support 临床辅助
Dietary 饮食
Future Dining 备用餐饮区
Inpatient Care 住院医护
Rehab. Services 康复服务
Support Services 辅助服务
Vertical Area 垂直区域

9

Fourth Floor Plan:
1. Head doctor
2. On-call
3. Staff facility
4. Nurse station
5. Conference
6. Dietary staff
7. Dietary
8. Staff dining
9. Light well
10. Atrium
11. Future dining

五层平面图：
1. 主管医生
2. 值班室
3. 员工设施
4. 护士站
5. 会议室
6. 员工餐饮
7. 餐饮
8. 员工就餐区
9. 进光孔
10. 中庭
11. 备用餐饮区

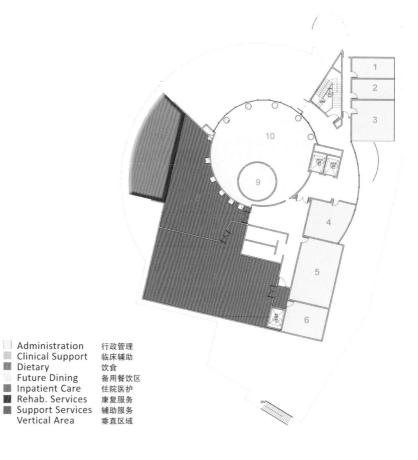

Administration — 行政管理
Clinical Support — 临床辅助
Dietary — 饮食
Future Dining — 备用餐饮区
Inpatient Care — 住院医护
Rehab. Services — 康复服务
Support Services — 辅助服务
Vertical Area — 垂直区域

190

9. Foyer and reception
10. Café, dining area
9. 大厅及接待处
10. 餐饮区

Children's Department and Work Therapy at the Institute for Rehabilitation of Republic of Slovenia, Ljubljana
斯洛文尼亚共和国卢布尔雅那康复研究所的儿科和工作治疗部

Architects: Dans Architects
Location: Ljubljana, Solvenia
Gross Floor Area: 4,569m²
Photographs: © Miran Kambic, Architect's archive
建筑师：丹斯建筑事务所
项目地点：斯洛文尼亚，卢布尔雅那
建筑面积：4569平方米
摄影师：©米兰·坎比克，建筑师设计档案

The whole Institute complex was designed between 1954 and 1962 by who envisioned low and elongated buildings, characteristic of Scandinavia, a pavilion set in a green environment. The new Children's Department and Work Therapy building with the children's unit on the ground floor and the vocational rehabilitation unit on the first floor keeps in with the pavilion-like design. It is linked with the existing complex by means of the interjacent middle part made of concrete and glass housing the lifts. The building itself has a rectangular floor plan with an internal atrium. The corridors and halls are well-lit and interesting. Along the sides of the atrium and the corridors, there are living spaces that open outwards into the exterior space with their wooden terraces. The large windows of the rooms and other spaces overlook the old park with the large trees, which have been

retained despite the park's having been reduced in area.

The simple architectural idea of a box in the middle of a park with a wooden atrium is based on the rational and functional plan. But this idea does not convey the whole identity of the architecture. The building tries to surpass the rigid functionality by using some fresh architectural ideas, for instance the wooden atrium, the playful graphics (by Slovenian artist Natasa Skusek) and the character of the exterior.

The building is built around the wooden atrium that enables the light to penetrate into the interior space. The wooden finish gives it a warmth and velvety softness. The users can use the atrium in a way it best serves their needs. The interior of the building is also graphically and sensorially exciting. The bold colours in the hallways with soft flooring serve as orientation around the building. The grey concrete walls of the core contrast the softness of the wood in the atrium, and the hygienic whiteness of the bathrooms is contrasted with bright coloured ceilings.

The building fits organically into the planted environment. The basic facade, made of fibre cement panels with a smooth finish, is white with large horizontal window apertures and a rhythm of vertical yellow accents made of corrugated panels. The volume of the building is perceived as a white canvas for projection of the graphics: the rhythmically placed yellow stripes, which should have been covered with an outer- green layer. The green layer would integrate the building into its environments and at the same time emphasise its poetics.

1. The new Children's Department and Work Therapy is a low and elongated buildings, a pavilion set in a green environment
2. Volume of the building is a white canvas for projection of the graphics: the rhythmically placed yellow stripes
3. The inner atrium with wooden finish is warm and velvety soft
4. Interjacent middle part that connects the old and the new building is made of concrete and glass
5. Vocational rehabilitation unit is located on the first floor
6. Grey concrete walls of the core contrast the softness of the wood in the atrium

1. 新建筑呈矮长形，是一个建在绿色环境中的亭阁式建筑
2. 建筑主体白色底衬托图案：黄色条纹有节奏地布局
3. 内部中庭由木质材料装饰，给人以温和、柔软之感
4. 居中部分连接了新老建筑，是由混凝土和玻璃构成
5. 职业病康复科位于建筑的第二层
6. 中心灰色混凝土墙体与中庭柔和的木质表面形成对比

7. Living spaces open into the wooden atrium
8. Users can use the atrium in a way it best serves their needs
7. 起居空间通向木质中庭
8. 使用者可以在中庭满足他们的需要

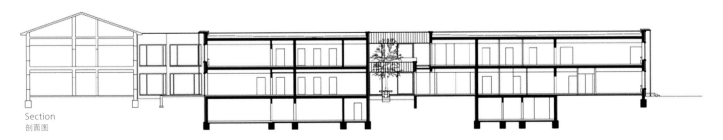

Section
剖面图

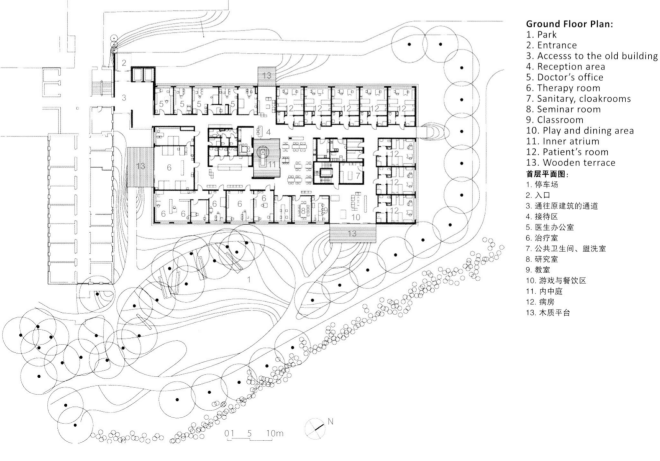

Ground Floor Plan:
1. Park
2. Entrance
3. Accesss to the old building
4. Reception area
5. Doctor's office
6. Therapy room
7. Sanitary, cloakrooms
8. Seminar room
9. Classroom
10. Play and dining area
11. Inner atrium
12. Patient's room
13. Wooden terrace

首层平面图：
1. 停车场
2. 入口
3. 通往原建筑的通道
4. 接待区
5. 医生办公室
6. 治疗室
7. 公共卫生间、盥洗室
8. 研究室
9. 教室
10. 游戏与餐饮区
11. 内中庭
12. 病房
13. 木质平台

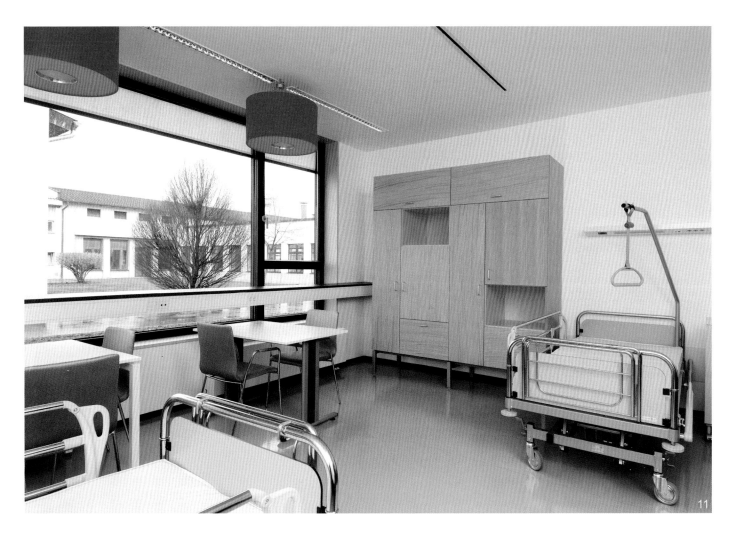

整个工程的设计工作从1954年到1962年间完成,方案包括斯堪的纳维亚半岛特色的低层、长形建筑,以及一个被绿色环绕的亭子。一楼的新儿科、工作治疗部与二楼的工作康复单元设计统一。混凝土中间结构将这两层建筑与原有建筑相连,电梯采用玻璃外壳。新工程的楼层平面为长方形,内有中庭。走廊和大厅光线充足,设计精巧。沿着中庭和走廊可以到达活动空间,活动空间的木质露台实现室内与室外的顺利过渡。通过窗户可以在室内俯瞰公园的景色。尽管公园面积缩减,园内的树木仍然得以保留。

通过科学的功能性规划,在公园中间建设一个内有木质中庭的方形建筑设计理念得以成功实现。但这个理念并没有充分地传达建筑的特色。设计师尝试通过创新的建筑设计理念,如木质中庭、活泼的图形(斯洛文尼亚艺术家娜塔莎·斯库塞克的作品)和特别的外形等,超越传统的功能性设计。

建筑内部的木质中庭设计大大加强室内空间的采光,木质建筑材料营造出温暖柔和的氛围。中庭空间可以服务用户的多种需求。建筑内部设计同样提供充分的感官享受。走廊选用柔软地面,配色大胆,可以起到指示作用。建筑中心的灰色混凝土墙壁与木质中庭的柔和质感形成对比,卫生间的洁白明亮则与鲜艳的彩色天花板相映成趣。

建筑与周围环境和谐相融。白色外墙的基础结构为表面光滑的纤维水泥板。墙上水平方向有明亮的大窗,竖直方向有黄色瓦楞纸板点缀。整个建筑就像一块白色的画布,不同的图形和色彩均在这里呈现:楼体规律的黄色条纹未来将会被一层绿色植物环绕。这绿色的元素不仅会模糊建筑与环境的界限,还会增添几分诗意。

9. Interior is graphically and sensorially exciting
10. Bold colours in the hallways with soft flooring serve as orientation around the building
11. Patients' rooms overlook the old park with the big trees with their large windows
12. Children's unit on the ground floor with a demonstration kitchen
9. 室内充满画面感,给人以愉悦之感
10. 走廊中大胆的色调与柔和的地板起到了定位作用
11. 通过病房巨大的窗体可以看到原建筑停车场内的大树
12. 儿童科位于建筑的首层,配备有示范厨房

Rehab-Hotel Sonnenpark Rust

朝阳公园康复酒店

Architects: skyline architekten
Location: Burgenland, Austria
Gross Floor Area: 7,500m²
Construction Area: 9,000m²
Project year: 2011
Photographs: © skyline architekten

建筑师：地平线建筑师事务所
项目地点：奥地利，布尔根兰
总面积：7500平方米
建筑面积：9000平方米
完成时间：2011年
摄影师：©地平线建筑师事务所

The project responds to the contour lines, the wooded slope, and the small buildings in the neighbourhood.

The two room levels of the upper floors are subdivided into 3 building sections set at an angle to each other with glazed connection. They are located on a broad single-storey base with therapy rooms, administration area, kitchen and dining hall. The rooms, orientated to the south, leave enough space to the wooded slope allowing good sunlight exposure, the rooms, orientated to the northwest, have an impressive view of the northern mountains as well as sunshine

from the west. The impregnated larch wood shuttering contributes to incorporation of the great cubage in the surrounding nature. Layout and orientation of the buildings allow a differentiated use of the southern garden on the wood side. The brick-clad therapy and ambulance building is directly adjacent to the park and the existing cardiovascular centre.

The low wing accommodates the curve of the street box, and then turns from the fall line of the hill to the south. The entrance hall passes through the valley side roomstract in its uppermost level, and ends in a terrace with panoramic views.

The terraced white building station defines a sheltered, south-facing courtyard, which also includes the recreational facilities are oriented. 160-bed rooms are divided into groups and stations, each unit has a coveredcommunity terrace.

The slope facilitates the clear separation of supply and disposal of the patient areasand provides the structure undisturbed by staggering visual links to the surroundingunspoilt countryside.

1. Side view of exterior
2. Courtyard with lawn and lounge area
3. View of exterior, detail
4. Building and its surrounding

1. 外观侧景
2. 庭院铺设草坪和休息区
3. 外观细部
4. 建筑及其周围环境

该工程设计融合当地树木茂盛的环境，倾斜的地势，以及周围的小型建筑。

酒店的上层结构分为三部分，以一定角度彼此相对，由玻璃通道相连。下面是宽阔的单层结构，包括治疗室、行政管理区、厨房和食堂。朝南的房间与树木的距离适中，确保采光充足。在朝西北的房间则可以将北方的山景尽收眼底，也可以在傍晚时分欣赏日落西山的景象。松木板使康复酒店与周围的自然环境更加和谐统一。

建筑的结构布局和朝向使得靠近树林的南侧花园可以服务多种功能。砖面外墙的治疗和救护车大楼距离公园和现有的心血管疾病中心很近。

低层建筑沿街而立，随着山坡的走势向南而落。入口大厅的最高处穿过河谷边缘，尽头是有开阔全景的露台。

康复酒店的白色大楼旁边还有一个朝南的院子。这里不仅遮风避雨，还提供多种娱乐设施。酒店的160个床位分成不同的组别和单元，每个单元都有独立的活动露台。

倾斜的地势将供给区和废物处理区明确分离，也使建筑能充分利用优越地势，尽情融入天然的乡村美景之中。

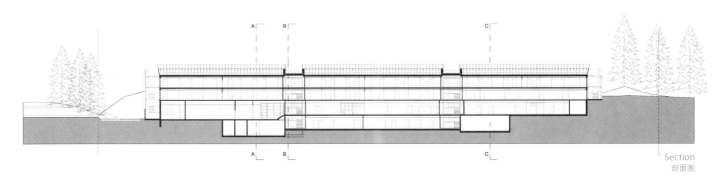

Section
剖面图

5. View of side balcony
6. Perspective view of corridor
5. 侧面阳台
6. 走廊透视图

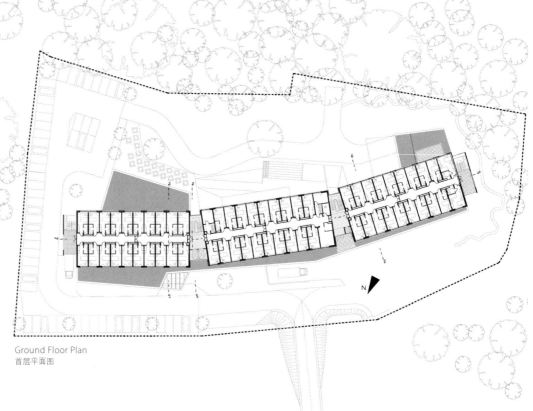

Ground Floor Plan
首层平面图

INDEX

Christian Kronaus + Erhard An-He Kinzelbach
ea@knowspace.eu

Dans Architects
dans@dans.si

d.A.p architecture & design
info@domenicoperronearchitetto.com

Dutch Health Architects (De Jong Gor temaker Algra)
Ronald.SchlundtBodien@egm.nl

Dutch Health Architects (EGM Architecten)
Ronald.SchlundtBodien@egm.nl

Enota
barbara.svetek@enota.si

GEA Arquitectos
c.torres@geasl.net

Gerhard Mitterber ger
oberwalder_zita@hotmail.com

索引

ippolito fleitz group / Peter Ippolito, Gunter Fleitz
info@ifgroup.org

Metropolisperu
info@metropolisperu.com

Mondaini Roscani Architetti Associati
studio@mondainiroscani.it

RTKL
info@rtkl.com

Seed Architects
a.burger@seedarchitects.nl

skyline architekten
office@skyline-architekten.at

weinbrenner.single.arabzadeh
kieser.ch@wsa-nt.de